# Illustrated Atlas of the Himalaya

# Illustrated Atlas of the Himalaya

(frontispiece) Machupuchhare Peak, located in central Nepal north of Pokhara Valley, is one of the few Himalayan summits still banned for climbing.

Copyright © 2006 by The University Press of Kentucky

Scholarly publisher for the Commonwealth, serving Bellarmine University, Berea College, Centre College of Kentucky, Eastern Kentucky University, The Filson Historical Society, Georgetown College, Kentucky Historical Society, Kentucky State University, Morehead State University, Murray State University, Northern Kentucky University, Transylvania University, University of Kentucky, University of Louisville, and Western Kentucky University. All rights reserved.

Editorial and Sales Offices: The University Press of Kentucky
663 South Limestone Street, Lexington, Kentucky 40508-4008
www.kentuckypress.com

10 09 08 07 06    5 4 3 2 1

Maps copyright © 2006 Julsun Pacheco
Photographs copyright © 2006 David Zurick

Cataloging-in-Publication Data is available from the Library of Congress.

The financial support of the following contributors is gratefully acknowledged:

American Alpine Club, Colorado
**American Geographical Society, New York**
Banff Centre, Alberta, Canada
Eastern Kentucky University, Richmond
National Science Foundation, Washington, D.C.

 This book is printed on acid-free recycled paper meeting the requirements of the American National Standard for Permanence in Paper for Printed Library Materials.

Design and typesetting by Julie Allred, BW&A Books, Inc.
Printed in China through Asia Pacific Offset.

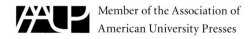 Member of the Association of American University Presses

# Contents

Maps vii

Tables and Charts ix

Acknowledgments xiii

Introduction xv

PART ONE  The Regional Setting 1

PART TWO  The Natural Environment 31

    Geology 33

    Climate 50

    Natural Hazards 60

PART THREE  Society 67

    Early Civilization 70

    Population 73

    Culture and Ethnicity 76

    Migration and Urbanization 81

    Transportation 85

    Communications 89

    Human Development 90

    Governance and Human Rights 96

PART FOUR  Resources and Conservation 99

    Agricultural Land 101

    Forests 112

    Minerals 120

    Water Resources 121

    Wildlife 130

    Future Trends 138

PART FIVE  Exploration and Travel 143

    Pilgrimages and Sacred Exploration 144

    Early Foreign Exploration 156

    The British Explorers 162

    The Mountain Climbers 170

    Trekkers and Modern Tourism 180

Sources of Illustrations 187

Bibliography 191

Subject and Name Index 195

Map Index 201

# Maps

Reference grid  x
Administrative districts  xi
Himalaya's place in the world  xii
Himalaya: regional setting  2
Himalaya Range  6–7
Himalaya: western sector  10
Himalaya: central sector  13
Himalaya: eastern sector  14
Major valleys and cities:
    Kathmandu Valley  20
    Kathmandu city  20
    Srinagar city  21
    Kashmir Valley  21
    Pokhara central valley  22
    Pokhara town  22
Bhutan valley settlements:
    Paro Valley  23
    Thimphu Valley  23
    Bumthang  23
Towns of the Himalaya:
    Gangtok  25
    Shimla town  25
    Dehra Dun  25
Tarai towns of Nepal:
    Biratnagar  26
    Nepalganj  26
Himalaya and Tibetan Plateau from the southeast  32
Drifting continents and formation of the Himalaya  34
Geology of Zanzkar and Indus Valley  35
Satellite map of the Himalaya  36–37
Geologic cross section  39
Northward drift of India  40
Generalized geology of the Himalaya  41

Himalayan geology:
    Kashmir Valley  43
    Kumaon region  43
    Nepal  43
Geology of Bhutan  47
Himalaya: mean temperature and climate  52
Ladakh and western Himalaya: precipitation:
    Ladakh winter  53
    Ladakh monsoon  53
    Western Himalaya winter  53
    Western Himalaya monsoon  53
Nepal: temperature and humidity  54
Seasonal climate system: air pressure and rainfall  57
Nepal precipitation patterns:
    Annual  59
    Highest 24 hour  59
    Monsoon  59
Himalaya: annual precipitation  59
Bhutan: annual precipitation  59
Himalaya: seismic hazard  60
Nepal: seismic hazard  60
Epicenters of major earthquakes  62
Major earthquakes of the Himalaya  62
Soil erosivity potential  63
Kathmandu Valley: liquefaction hazard  64
Pokhara Valley: subsidence hazard  65
Himalaya: cultural regions  69
Major Himalayan trade routes  70
Himalaya: population density  73
Himalaya: population growth  74
Himalaya: ethnic groups  75
Nepal: ethnic groups  76
Nepal: languages  76

Religions in Nepal: Hindu, Muslim, Buddhist  77
Indian Himalaya: tribal population  81
Indian Himalaya: urban population  81
Towns in Indian Himalaya  83
Towns in eastern Himalaya  84
Himalayan roads  86
Nepal: road network  86
Nepal: airports  88
Communications in Bhutan  89
Nepal: landless and marginal farm households  91
Literacy rate  92
Bhutan: educational facilities  92
Nepal: access to potable water  94
Kathmandu Valley: water quality  94
Bhutan: health facilities  94
Himalaya: landscape regions  100
Himalaya: elevation zones  100
Himalaya: land use  102
Himalaya: soil classification  103
Himalaya: cropland distribution  104
Himalaya: population and farmland  108
Himalaya: annual farmland change  109
Nepal: cultivated area  109
Nepal: irrigated area  109
Nepal: sloping terraced area  109
Nepal: grassland  110
Nepal: chemical fertilizer use (1990s)  111
Himalaya: forest cover  113
Nepal: per capita forest area  114
Forest change in the tarai districts (1977–1994)  114
Himalaya: population and forest area  116
Bhutan: fuelwood deficit areas  115
Himalaya: forest cover change  117
Nepal and Bhutan: mineral deposits  120
Main rivers of the Himalaya  122

Changing course of Kosi River, Nepal  123
Hydroelectric projects in the Bhagirathi River basin  125
Nepal: hydropower development  126
Glaciers, glacial lakes, and watersheds of Nepal and Bhutan  127
Royal Chitwan National Park  131
Royal Chitwan National Park and surrounding village population density  131
Himalayan national parklands  135
Nepal: protected areas  137
Bhutan: protected areas  137
Tarai arc conservation landscape  138
Ancient Hindu holy places  146
Char Dham: Hindu pilgrimage to Himalayan temples  152
Pilgrimage route to Muktinath  154
Explorations by Chinese travelers  157
D'Anville map  158
Early exploration by Europeans  160
Early explorations: regional setting  160
Sven Hedin's map of major Himalayan river sources  161
Early explorers' routes in east-central Himalaya  162
Early explorers' routes in western Himalaya  164
Routes of Indian and Tibetan explorers  167
Bhutan exploration: journeys of John Claude White  171
First ascent routes of 8,000-meter peaks:
    Nanga Parbat  172
    K2  172
    Everest  173
    Cho Oyu  173
    Makalu  173
Himalaya trekking and mountaineering regions:
    Overview  178
    Annapurna  178
    Everest  179
    Zanzkar and Ladakh  179

# Tables and Charts

Jammu and Kashmir fact file  4
Himachal Pradesh fact file  5
Uttaranchal fact file  5
Nepal fact file  15
Area of Indian Himalaya states and territories  15
Administrative regions of Indian Himalaya  15
Sikkim fact file  16
Bhutan fact file  17
Arunachal Pradesh fact file  18
Geoecology of the western mountains  45
Nepal: temperature and precipitation  55
Orographic precipitation  57
Indian Himalaya: rural population (1991)  83
Number of towns by population  83
Nepal: urbanization (1952–2001)  84
Nepal: tourist arrivals  91

Nepal: percent of population with access to drinking water (1985–1999)  95
Indian Himalaya: land use  104
Nepal: land use (1999)  105
Indian Himalaya: agricultural crops  106
Nepal: change in population and cultivated area (1975–1999)  106
Nepal: livestock and grazing area  110
Nepal: population growth and fuelwood consumption (1981–1997)  115
Nepal: export of forest products (1975–1998)  118
Stream flow in three rivers  123
Indian Himalaya: percent of households with access to drinking water and electricity  126
Bhutan: protected areas  137
Himalaya: protected area and biodiversity  138

# REFERENCE GRID

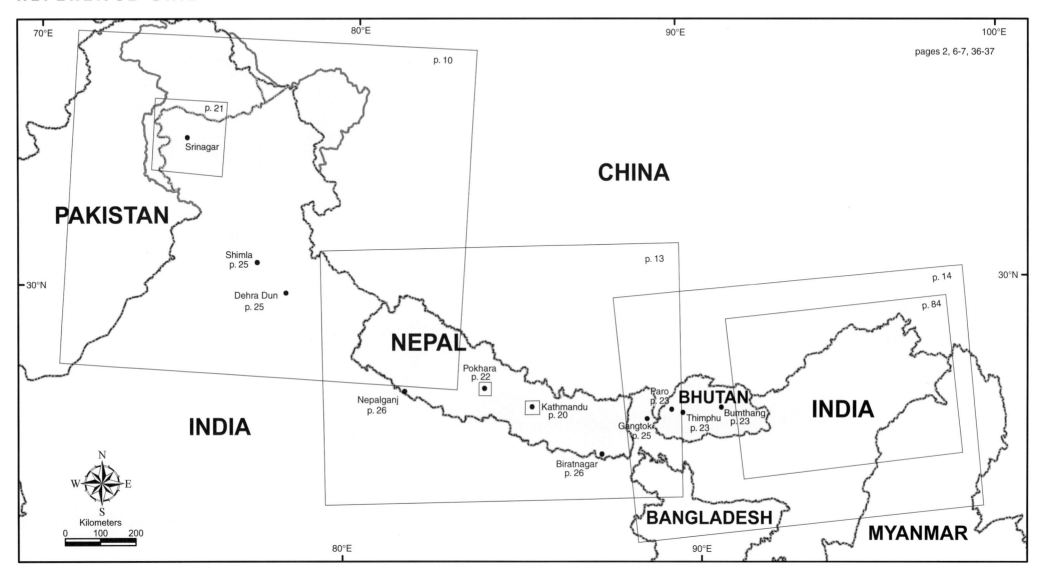

pages 2, 6-7, 36-37

**Physical Features**
- )( Pass
- K2 (8611) ▲ Mountain peak
- Lahaul Valley Physical feature
- Road
- River
- Lake

**Urban Areas**
- **Srinagar** ◉ Major city
- **Anantnag** ◎ Secondary city
- **Chamba** • Town
- **Arakot** • Settlement area
- **Chandigarh** ◉ Selected city

**Political Boundaries**
- ––––– Disputed boundary
- ––––– International boundary
- ––––– Line of control

**Elevation (meters)**

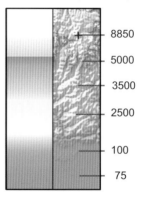

>> ADMINISTRATIVE DISTRICTS

The Himalayan states are divided into 149 administrative districts for purposes of local governance.

HIMALAYA'S PLACE
IN THE WORLD

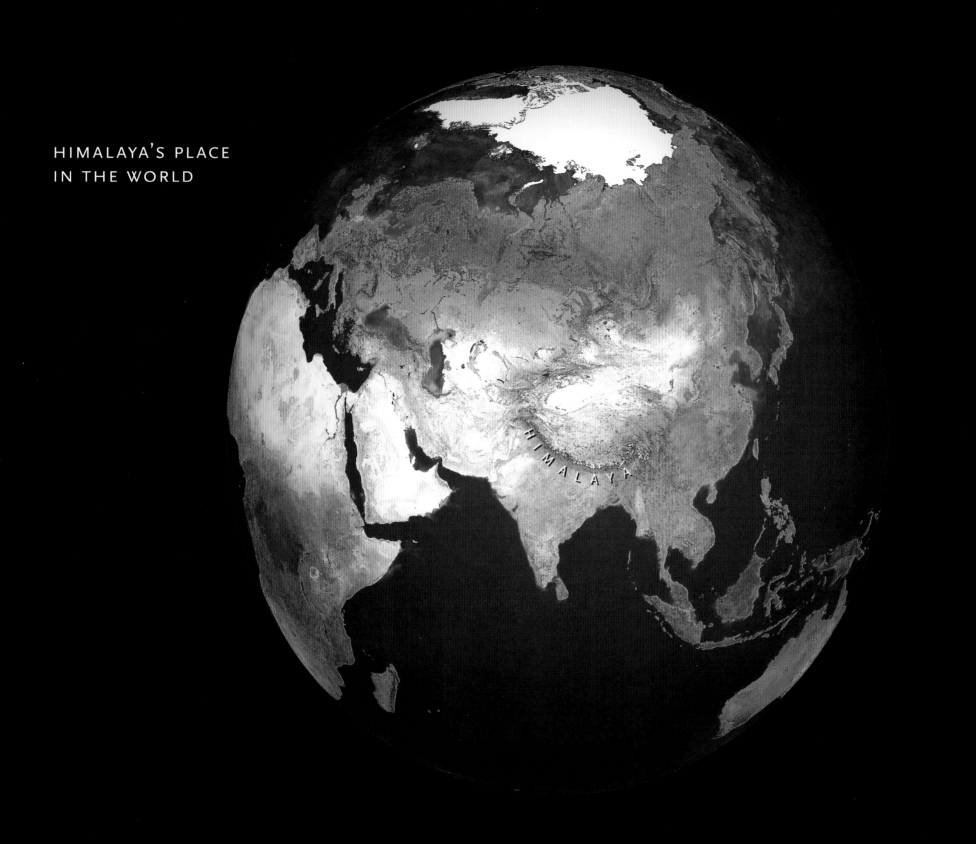

# Acknowledgments

This atlas is the culmination of many years of fieldwork in the Himalaya and cartographic work in Kentucky. In the former case, the mountains were traversed on foot, on horseback, by jeep, and by bus. In the latter case, the maps were rendered digitally in a computerized geographic information system. On both counts, the authors acknowledge and appreciate the assistance of numerous individuals and agencies.

Data compilation in the Himalaya for purposes of the atlas was facilitated by the International Centre for Integrated Mountain Development (ICIMOD). The current volume springs from an earlier document prepared with the assistance of ICIMOD staff members Basanta Shrestha and Birendra Bajracharya, utilizing some of ICIMOD's extensive digital data for the Himalaya. We are indebted to ICIMOD both for providing technical assistance and for producing that prior version of the atlas. Thanks also go to Dr. Harka Gurung, a world authority on the Himalaya, for reviewing early versions of the atlas and making significant editorial contributions. For field assistance in the Himalaya, the authors thank the following persons: Ibrahim Chapri (Kashmir); Tsering Phuntsok (Ladakh); Parvez Bhat (Ladakh and Zanzkar); Dorje Rinchen (Zanzkar); Sub-Inspector Singh (Lahaul and Spiti); Dilli Raj Joshi (Garhwal); Birendra Bajracharya, Prakash Pathak, Chris Monson, Mani Lama, Suresh Khadkha, Basanta Shrestha, and Rajendra Shrestha (Nepal); and Chencho Tshering, Karma Dorji, and Dawa Zangpo (Bhutan). A number of governmental and nongovernmental agencies assisted in the completion of fieldwork and the compilation of data, including the following: King Mahendra Trust for Nature, World Wildlife Fund–Nepal, Friends of Chipko, Ministry of Population and Environment–Nepal, Ministry of Tourism and Civil Aviation–Nepal, Wadia Institute of Geology, G. B. Pant Institute of Himalayan Environment and Development, Bhutan Trust Fund for Nature, Tribhuvan University, Map Point, Planning Commission–Nepal, Royal Government of Bhutan, and Department of National Parks and Wildlife Conservation–Nepal.

Financial support for Himalayan field studies over the years and for production of the atlas was provided by generous grants from the National Science Foundation, the American Geographical Society, American Alpine Club, Banff Centre, and Eastern Kentucky University. Eastern Kentucky University also provided excellent institutional support, including office space, leaves of absence, and a sabbatical research award for David Zurick. In support of his cartography, Julsun Pacheco thanks Gyula Pauer, Ev Wingert, and especially Richard Ulack, to whom the maps in the atlas are dedicated.

Special thanks go to Pradyumna Karan for his mentorship and unflagging support.

Julsun Pacheco is grateful to his wife, Cindy, and daughter, Natalie, for putting up with his disappearances into the digital cave in his house, and to his parents in the Philippines, who thought that he only knew how to build computers and were especially proud to learn that he could do something useful like make maps. David Zurick thanks his wife, Jennifer, for patiently enduring his absences while he was away in the mountains and, especially, for accompanying him on some of his fieldwork journeys.

Finally, we wish to thank Steve Wrinn, director of the University Press of Kentucky, for his enthusiastic embrace of this project, and all the Press staff who have helped turn our manuscript into this book.

# Introduction

The political geography of the Himalaya is contested in many places, with China asserting its claim over territory occupied by India and Bhutan, and India asserting its claim over territory occupied by Pakistan and vice versa. The boundary lines that appear on the maps in this atlas represent the most widely accepted designations of country borders; in some cases, again by convention, they appear as something other than a recognized boundary line (for example, the disputed boundary between Pakistan and India in Kashmir is depicted as a "line of control"). Such disagreements about boundary placements suggest the somewhat fluid nature of Himalayan geopolitics, which historically includes diverse relations among the neighboring mountain countries. The administrative divisions of territory within the individual Himalayan countries are also subject to change, most notably with the emergence of Uttaranchal in 2000 as a new mountain state of India, but also with countless smaller efforts to redraw district and settlement boundaries all across the mountains.

Similarly, the regional geography of the Himalaya is subject to interpretation. As an alpine system, the Himalaya is connected to the much larger South Asian mountain rim land, which extends from the Hindu Kush Range in Afghanistan to the Hengduan Range of Yunnan and Sichuan provinces in China. In this long tectonic reach lie several of earth's greatest mountains, including the Hindu Kush in Afghanistan, the Karakoram in Pakistan, the Pamir in Tajikistan, the Zanzkar in India, and the Kunlun in Tibet (Xizang) and China, as well as the Himalaya. The last stretches from northwestern India across the subcontinent to northeastern India, capturing along the way the kingdoms of Nepal and Bhutan. No sharp geographic divisions actually exist in the landscape between the ranges, and people commonly group them when speaking of the Himalaya. We use in this atlas a fairly conventional but restrictive geographic interpretation of the Himalaya, which delimits the mountains bounded in the west by the Indus River, in the east by the Brahmaputra River, and in the north and south by, respectively, the Tibetan Plateau and the plains of India.

# PART ONE The Regional Setting

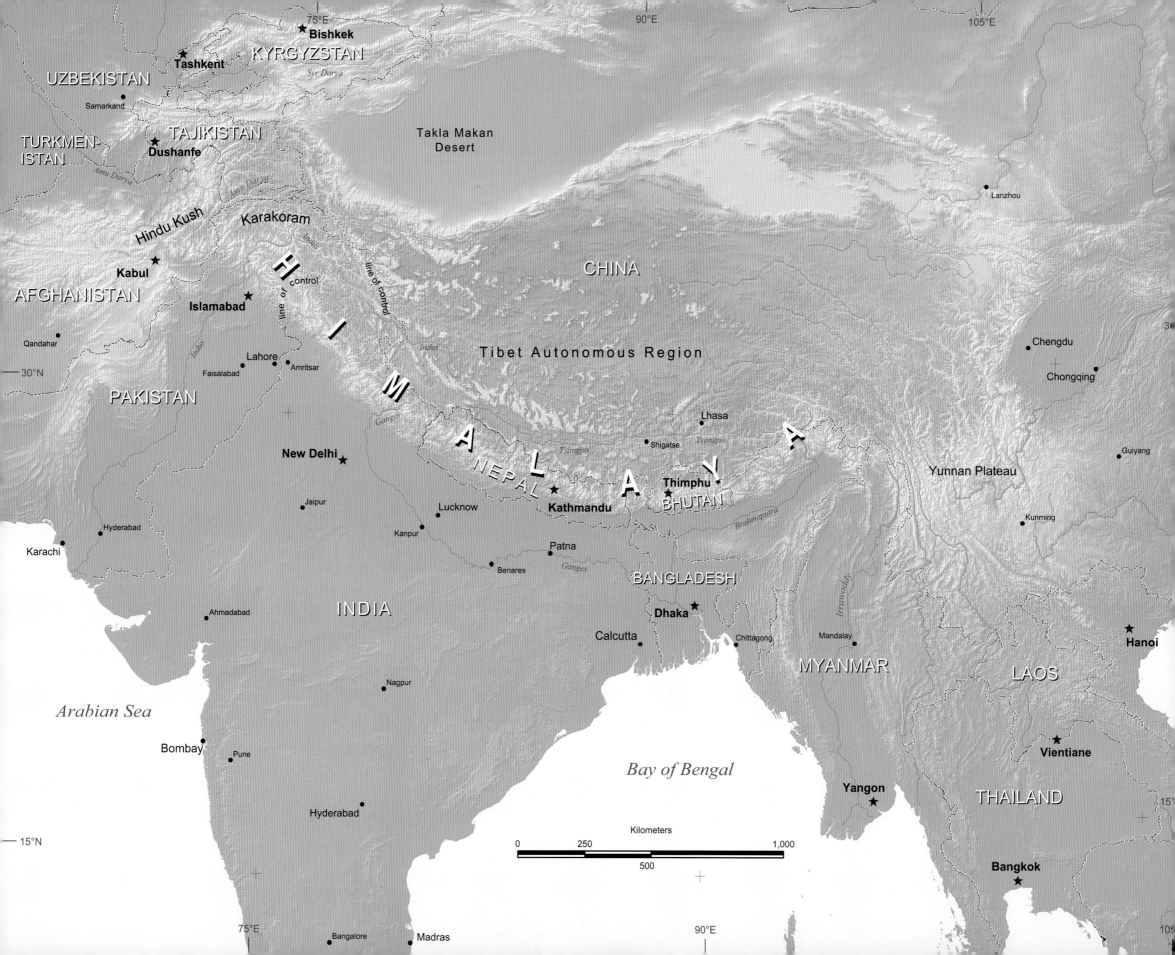

# *Hima' al-aya:* abode of snow (Sanskrit)

The slow, inexorable drift northward of the South Asian continent, beginning about 130 million years ago in the Cretaceous period and continuing into the present day, resulted in the collision of the Indian and Asian continental plates, uplifting huge sections of old, compressed sea floor and creating an extraordinary range of mountains—the Himalaya, which today stands high above all other places on earth. Evidence of this cataclysmic event, and of the mountains' 60 million–year–old roots, can be found in the landscape, among the twisted strata of exposed rock that show the enormous pressures of buckling and folding, and in the fossilized brachiopods, corals, and skeletal fish that are trapped at the foot of glaciers several kilometers high. These demonstrate the oceanic origins of the mountains; seashells are found even near the summit of Mount Everest. Among the remnants of the ancient Tethys Sea, which stood between India and Asia during much of their geologic convergence, are the ammonites, whose spiraling shape represents for both Hindus and Buddhists the "cosmic mandala." They are especially esteemed by pilgrims as talismans for the power that comes from the curious mix of geology and mythology.

From the southern plains, the mountains appear as an impossible line of white peaks stretching to the horizon, like a physical rampart separating entire worlds. Peoples living both north and south of the Himalaya believe it to be the abode of the gods, and the great deities of Tibet and India are thought to reside there and to intermingle in one another's affairs. For this reason, the Himalaya is held to be holy by native people and is believed to be animated with divinity. The religious view ascribes the celestial heights of the mountains to the *axis mundi*, which symbolically links the spiritual and secular realms in a conjunction of heaven and earth and populates the mountainous landscape with mythical creatures, treasure places, and monastic settlements, thus creating auspicious settings filled with inexplicable power. Magic and belief give people an enduring place in the rugged and magisterial terrain of the Himalaya.

A geographic view of the mountains celebrates its stunning physical qualities. When the Karakoram peaks are included, the Himalaya contains all fourteen of earth's summits over 8,000 meters, including Mount Everest, which at 8,850 meters is the world's highest mountain. The deepest canyon in the world—the Kali Gandaki Gorge—is also found there, as are numerous other world-class valleys, including some so remote,

Khumbu Himal, Nepal.

⌃ (part opener) The village of Siklis, located on the southern slopes of Annapurna, is inhabited by the Gurung people.

« HIMALAYA: REGIONAL SETTING

The South Asia highlands extend from Afghanistan to Myanmar and encompass the Hindu Kush, Karakoram, and Himalaya ranges (the HKH region). To the north of the HKH region is high Central Asia, which includes the Pamir mountain states of Tajikistan and Kyrgyzstan, as well as the massive Tibetan Plateau. This entire area is the loftiest place on earth. South of the mountains are the peninsula of India and the adjoining countries of Pakistan and Bangladesh. All the countries of continental South Asia are geographically tied to the Himalaya, either directly by territory located in the mountains or indirectly because the alpine system fundamentally interacts with the land and water systems within their political reach.

« Artistic rendering of central Himalaya. (Painting by Shyam Tamang Lama; used with permission of the artist)

Indus River in Baltistan, Pakistan, forming the western boundary of the Himalaya.

### JAMMU AND KASHMIR FACT FILE

| | |
|---|---|
| Form of government: | Federal republic (state of Indian Union) |
| Capital: | Jammu (winter), Srinagar (summer) |
| Area: | 222,236 square kilometers |
| Population: | 7,718,700 |
| Population density: | 34.7 persons/square kilometer |
| Life expectancy: | No data |
| Infant mortality: | No data |
| Official or principal languages: | Hindi, Kashmiri, Dogri-Kangri |
| Literacy rate: | No data |
| Religions: | Muslim 74%, Hindu 23%, Buddhist 3% |
| Currency: | Indian rupee |
| GNP per capita: | No data |
| Climate: | Temperate to alpine |
| Highest point: | Nun, 7,135 meters |

such as the Tsangpo Falls Gorge, that they have only recently been discovered. In its regional setting, the Himalaya is the topographic showpiece of a huge highland area stretching from Kyrgyzstan to Myanmar and encompassing the Pamir, Hindu Kush, and Karakoram ranges, as well as the Himalaya proper. These ranges coalesce into a contiguous ridge of folds and upthrusts along a 4,000-kilometer crescent. In a strict geographic sense, the Himalaya proper is more narrowly defined. It occupies the territory between 75 and 95 degrees east longitude and 27 and 35 degrees north latitude, extending from the Indus River in the west to the Brahmaputra River in the east, anchored by the summits of 8,125-meter Nanga Parbat and 7,756-meter Namche Barwa, respectively. It encompasses a geographic relief from the rolling high plateau of Tibet in the north to the outer foothills of the Gangetic Plain in the south. According to this definition, the Himalaya is about 2,700 kilometers long, contains 600,000 square kilometers, and has a population of more than 47 million people.

With its huge tract of space and immense geographic proportions, the Himalaya is distinguished by a great variety of climate, terrain, and human culture. The western regions of Ladakh and Zanzkar are dry, and north of the main summits is desert; temperatures are cold for much of the year. The eastern edge of the Himalaya, in the watershed of the Brahmaputra River where it disgorges from the Kham highlands in Tibet, is subtropical and one of the wettest spots on earth, with recorded precipitation in excess of 14 meters a year. From west to east, north to south, there are enormous gradients of both precipitation and temperature. To the extent that the regional climate influences the distribution of natural vegetation, there is also great complexity in Himalayan plant life. The eastern region of Bhutan and the upper Assam Valley are thought to be biodiversity hotspots, with a tremendous variety of plants and animals, many of which are found nowhere else in the world.

More striking, perhaps, than the horizontal space occupied by the Himalaya is the vertical aspect of the mountains. Changes in elevation result in the formation of different climatic and vegetation zones along the flanks of the mountains. On average, the temperature difference with altitude change is 6.4 degrees Celsius per kilometer; with elevation changes of almost 8,000 meters possible, the temperature gradient in the heights of the Himalaya is comparable to what would be encountered when traveling from the tropical latitudes to the polar ice fields. Add to the vertical gradient the fact that slope aspects determine solarization and windward versus leeward localities witness extreme differences in precipitation, and the possibilities for natural diversity over short distances are great. This complex environmental model confuses most attempts to describe the Himalayan environment in monolithic terms. On the ground, however, the diversity is easily seen: The terrain falls away from the icy peaks in terraces and deep green valleys, rolling across ridge upon ridge in cascading descents before spilling onto the lowland plains. Water follows the contours of the land, gushing from glacial lakes and snowmelt in torrents, cutting shadowy gorges, and nourishing agricultural fields carved into the sides of the mountains. The surface of the land, in response to the diverse parameters imposed by terrain and climate, is a bewildering patchwork of glaciers, forests, pastureland, rivers, and farms. This natural diversity is accentuated by the Himalaya's complex cultural geography, which weds human society to the physical circumstances of the mountains.

Three major civilizations converge in the Himalaya: the Buddhist monastic tradition from Tibet, the Islamic societies

Mount Everest, at 8,850 meters, is the highest mountain on earth.

## HIMACHAL PRADESH FACT FILE

| | |
|---|---|
| Form of government: | Federal republic (state of Indian Union) |
| Capital: | Shimla |
| Area: | 55,673 square kilometers |
| Population: | 6,177,248 |
| Population density: | 111 persons/square kilometer |
| Life expectancy: | 60 years |
| Infant mortality (per 1,000): | 121.92 |
| Official or principal languages: | Pahari, Hindi |
| Literacy rate: | 63.50% |
| Religions: | Hindu 96%, Muslim 1.6%, Buddhist 1.2% |
| Currency: | Indian rupee |
| GNP per capita: | US$375 |
| Climate: | Temperate to alpine; monsoon |
| Highest point: | Leo Pargyal, 6,790 meters |

## UTTARANCHAL FACT FILE

| | |
|---|---|
| Form of government: | Federal republic (state of Indian Union) |
| Capital: | Dehra Dun (interim) |
| Area: | 55,845 square kilometers |
| Population: | 7,000,000 |
| Population density: | 125 persons/square kilometer |
| Life expectancy: | 55 years |
| Infant mortality (per 1,000): | 113 |
| Official or principal languages: | Pahari, Garhwali, Hindi |
| Literacy rate: | 58% |
| Religion: | Hindu |
| Currency: | Indian rupee |
| GNP per capita: | US$240 |
| Climate: | Temperate to alpine |
| Highest point: | Nanda Devi, 7,817 meters |

The Regional Setting 5

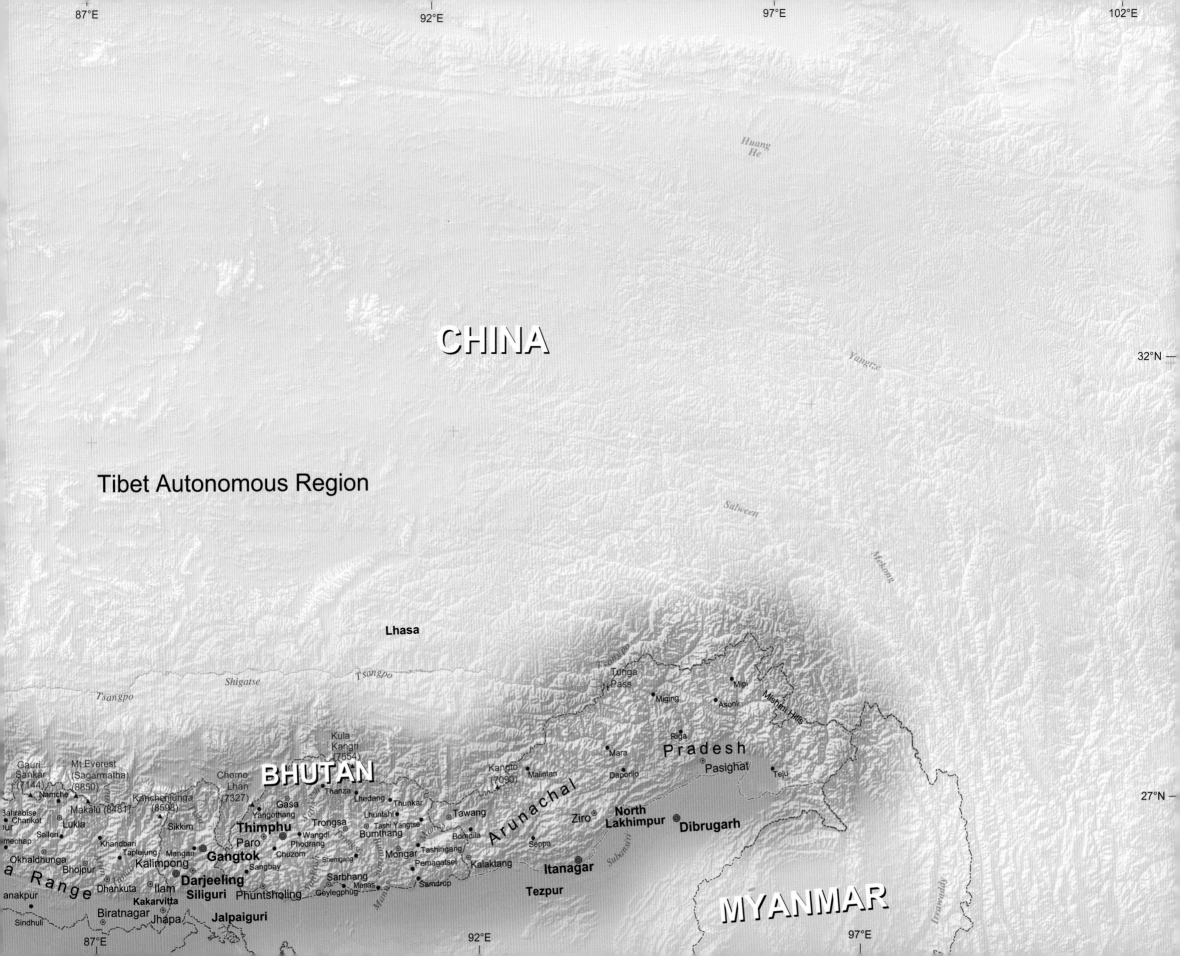

Ladakh, in the Indus Valley near the contested boundary between India and Pakistan.

« (previous pages) HIMALAYA RANGE

The Himalaya extends northwest to southeast in a 2,600-kilometer crescent of highlands between the Indus and Brahmaputra rivers. The mountain range contains the world's highest peaks and has over 47 million inhabitants.

in the western region, and the Indic-Hindu culture from India. These societies overlay the numerous tribal settings, from the pastoral and seminomadic communities in the trans-Himalayan valleys to the slash-and-burn farmers in the lowland subtropical forests, so that an extraordinary assemblage of local cultures results from the blending of traditions. The kaleidoscopic human face of the Himalaya attests to the unique and compelling ways in which people have adapted to the mountain world. Anthropologists provide detailed ethnographies of Himalayan peoples, showing them to be resilient and tenacious in the midst of the demands imposed by local environmental conditions. It is less clear, however, how the traditional lifestyles will help people navigate the future, where the pressures of limited resources will combine with the insatiable demands of globalization.

In modern times, the peoples of the Himalaya have been forced to accommodate the needs of much larger societies, which conventionally view the mountains as sovereign terri-

Nakho village, western Himalaya.

« HIMALAYA: WESTERN SECTOR

The Indus River marks the western boundary of the Himalayan region, separating that range from the adjoining Karakoram. The Indus first flows northwest from its source near Mount Kailas through the high desert of Ladakh before veering south through the High Himalaya near Nanga Parbat and then emptying onto the plains of Pakistan. Within the bends of the Indus is a small sliver of land belonging to Pakistan, as well as the contested region of Kashmir. To the east, the western Himalaya encompasses the Indian states of Himachal Pradesh and Uttaranchal, up to the border of Nepal. An enormous change of landscape occurs when crossing the high passes from south to the north; the greenery and rugged terrain of the Low and High Himalaya give way to the rocky plateaus of the trans-Himalayan zone in Ladakh and Spiti.

Shermatang village, Yolmo, Nepal.

tory and important resource frontiers. The Asian countries bordering the Himalaya have carved the mountainous territory into respective political possessions, albeit with great uncertainty in some places. Kashmir, for example, is contested by India and Pakistan, whereas many Kashmiri people would prefer an independent state. India and China differ over the ownership of an inhospitable stretch of cold desert north of Ladakh known as the Aksai Chin. India initially claimed the land and put it on its maps but discovered in 1958 that China had already built roads through the region. Tiny Sikkim, wedged between Nepal and Bhutan, was an independent kingdom until it became the twenty-second state of India in 1975. Chinese maps include about 300 square kilometers of territory that belongs to Bhutan, but the matter is not pursued because China wishes to maintain cordial relations with Bhutan. Much of the eastern region of the Himalaya, now occupied by the

The Regional Setting 11

Paro Dzong, Bhutan.

## HIMALAYA: CENTRAL SECTOR

Nepal and Sikkim compose the central Himalaya, where the highest peaks of the range are found. The monsoon climate dominates this region. High population densities and poverty are common, and some of the most severe environmental degradation occurs here. Much of the interior region of the central Himalaya remains roadless, and barring the few airstrips capable of handling small aircraft, the mountainous areas are accessible only by foot. The narrow strip of land along the Nepal-India border, known as the *tarai,* is one of the most rapidly developing zones in the entire Himalaya.

Indian state of Arunachal Pradesh, is claimed by China. These contested divisions of Himalayan territory, long a part of the region's political history, continue in the present day; most recently, the Himalayan state of Uttaranchal was formed in 2000 by assembling thirteen mountain districts of Uttar Pradesh into India's twenty-seventh state.

The demarcation of political areas in the mountains results in the boundaries that commonly appear on Himalayan maps. In the western region, the Pakistan districts of Mansehra, Abbottabad, and Kohistan share the Indus watershed and lie in the shadow of Nanga Parbat. This part of Pakistan is known as Azad Kashmir, meaning "free Kashmir," and reflects Pakistan's viewpoint that much of the rest of Kashmir is illegally occupied by India. Arcing westward, the Himalaya leaves Pakistan sovereign territory and enters the Indian state of Jammu and Kashmir. It is a tumultuous zone, with transparent borders that are poorly enforced despite a heavy military presence by both Pakistan and India. More than 200,000 square kilometers

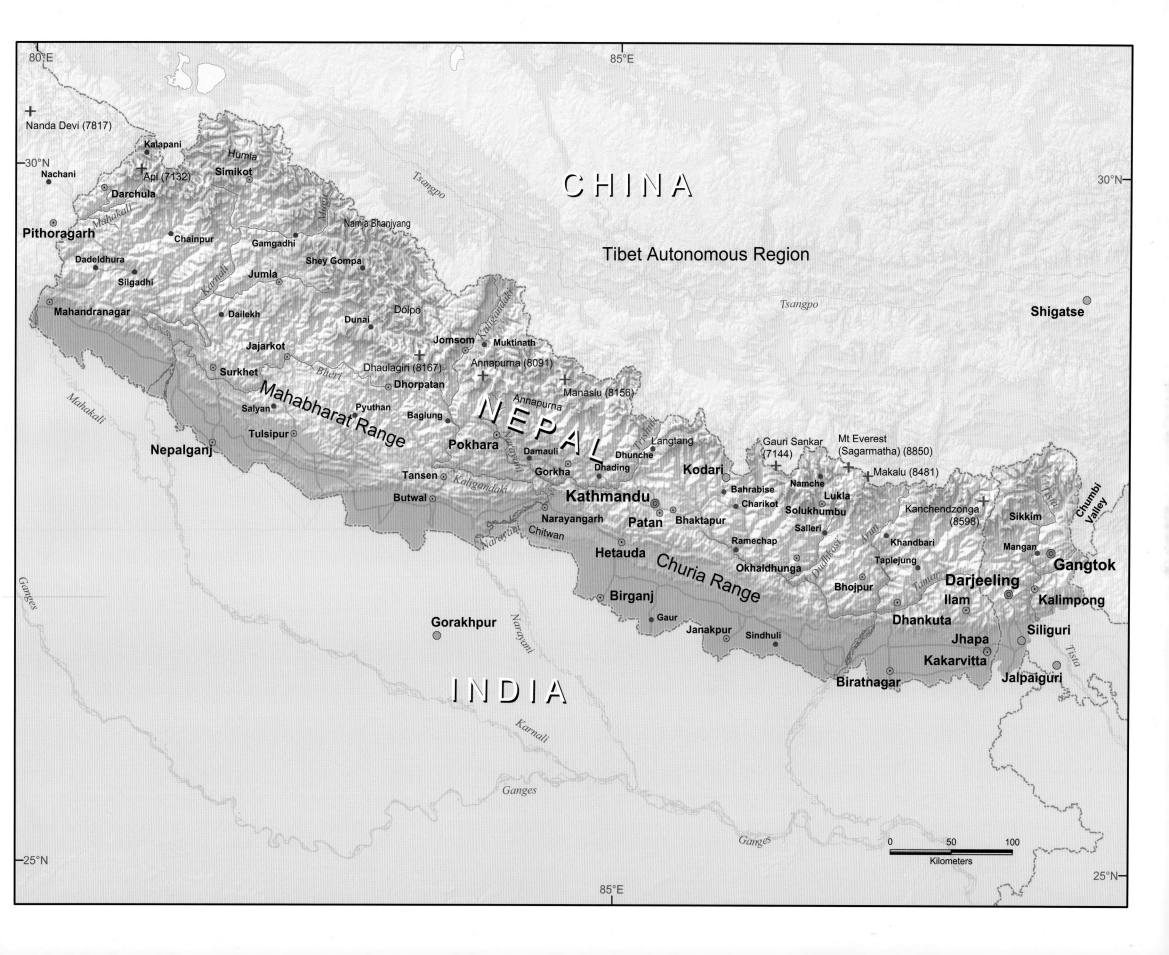

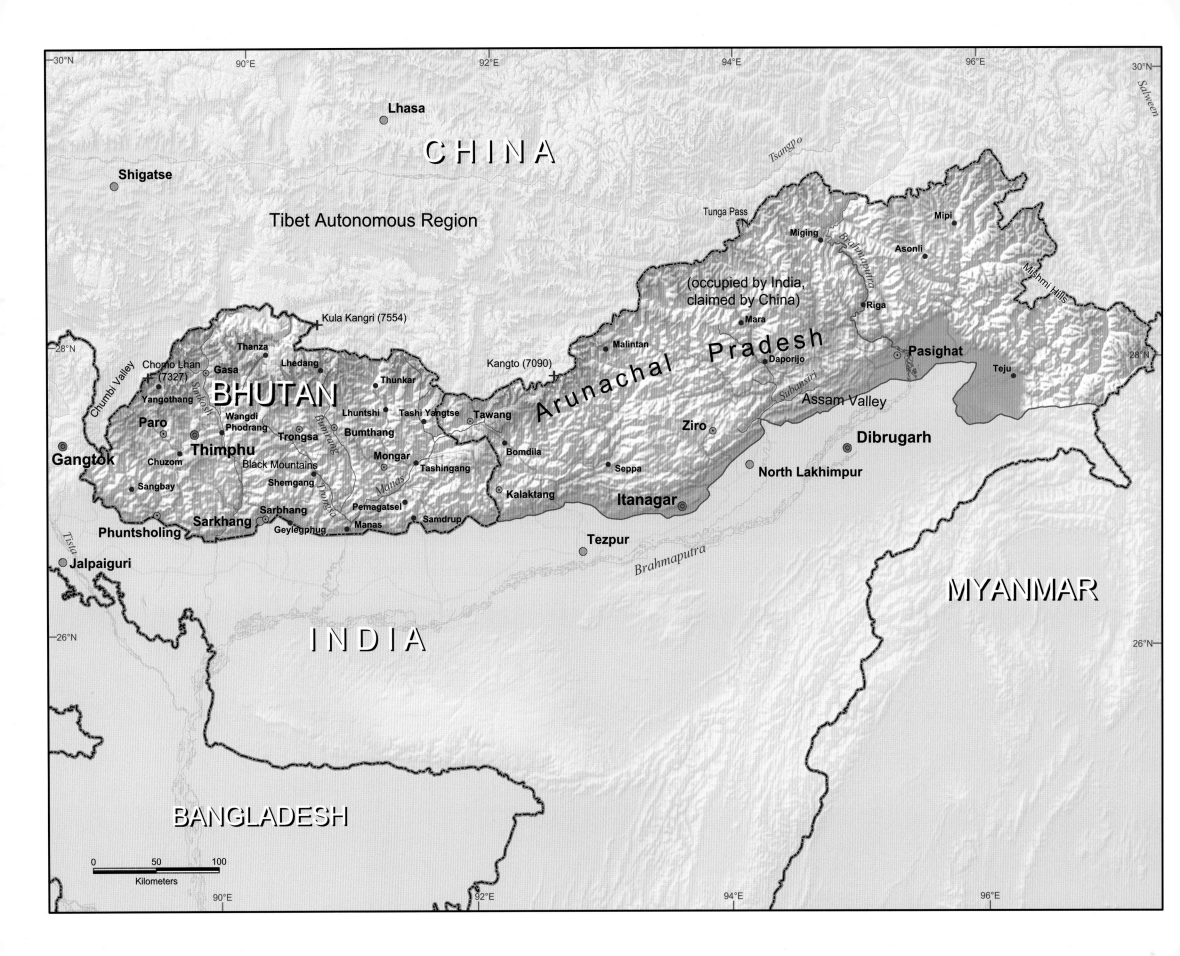

where it shifts southward from the Tibetan Plateau (where the river is known as the Yarlung-Tsangpo) at 95 degrees east longitude. There, it cuts a deep gorge in the Himalaya before emptying into the Assam Valley near the river town of Dibrugarh.

In total, the Himalaya encompasses a remarkable mosaic of landscapes and cultures, organized in part according to the political territories described earlier, but also in accordance with long-standing heritages that predate the modern nation-states. The naming of Himalayan places, which gives rise to the titles that appear on maps, also highlights the juxtaposition of old settlement titles and modern names derived from contemporary nation building. In some cases, the old and new names are used interchangeably, such as Everest and Sagarmatha, but most often, the ancient toponyms, which hearken back to dialects that may no longer exist, have given way to their modern equivalents in the national language or in English. The word *dzong,* for example, which still appears on Himalayan maps in reference to distinct places from Ladakh in the west to Bhutan in the east, reflects an early Tibetan dialect and refers to a fortress or a fortress-like settlement. Commonly, a prefix locates the fortress in a specific geographic locality. For example, Baragaon Dzong in Nepal refers to a valley of twelve fortress-like villages north of the Annapurna Range, while Paro Dzong in Bhutan refers to a historical center of the kingdom. The term *dzong,* however, is rarely used outside Bhutan on new maps, with a few Nepalese exceptions. The names of mountain summits, likewise, commonly appear in multiple ways. Thus, the world knows the highest mountain on earth as Mount Everest, which commemorates the British surveyor Sir George Everest, who first measured its height, but the summit, located in northeastern Nepal, is known by the Nepalese as Sagarmatha, which means "mother goddess," and by the Tibetans as Chomolungma, or "goddess mother of the earth."

In a general terrain overview, the Himalaya exhibits a tiered profile, with north-south stepwise zones of elevation corresponding to the geologic structures. North of the high peaks are remote sections of arid plateaus and valleys. In Ladakh, the

Tharu tribal woman harvesting wheat in a tectonic (dun) valley in the outer foothills zone. The Siwalik hills loom to the north in the background.

### BHUTAN FACT FILE

| | |
|---|---|
| Form of government: | Monarchy |
| Capital: | Thimphu |
| Area: | 47,000 square kilometers |
| Population: | 1,951,965 |
| Population density: | 41.5 persons/square kilometer |
| Life expectancy: | 52.8 years |
| Infant mortality (per 1,000): | 109.3 |
| Official language: | Dzongkha |
| Literacy rate: | 41.1% |
| Religions: | Lamaistic Buddhist 75%, Hindu 25% |
| Currency: | Ngultrum |
| GNP per capita: | US$420 |
| Climate: | Tropical to alpine; monsoon |
| Highest point: | Kula Kangri, 7,554 meters |

### ARUNACHAL PRADESH FACT FILE

| | |
|---|---|
| Form of government: | Federal republic (state of Indian Union) |
| Capital: | Itanagar |
| Area: | 83,743 square kilometers |
| Population: | 1,091,117 |
| Population density: | 13 persons/square kilometer |
| Life expectancy: | 44 years |
| Infant mortality: | No data |
| Official or principal languages: | Dafla, Abori, Mishmi |
| Literacy rate: | No data |
| Religions: | Hindu 37%, Buddhist 13%, Tribal 37%, Other 13% |
| Currency: | Indian rupee |
| GNP per capita: | US$305 |
| Climate: | Tropical to alpine |
| Highest point: | Kangto, 7,090 meters |

Thak Khola Gorge, Nepal.

Tibetan Plateau extends for several hundred kilometers and shapes a 60,000 square kilometer landscape of arid steppes and valleys, similar to that found farther north in Tibet. Much of this area is windy and cold, possessing a stark beauty but containing little water and few places suitable for human settlement. Other minor extensions of the plateau occur in isolated regions of Nepal, notably in Dolpo and Mustang, which lie north of 8,167-meter Dhaulagiri and 8,091-meter Annapurna, respectively; in Sikkim along the upper Tistha River; and in Bhutan north of 7,320-meter Kula Kangri.

The remote plateau region lies in the rain shadow of the great Himalayan peaks and is therefore arid, with surface water originating mainly from the melting of snowfields and glaciers. The terrain and isolation make communication and travel across this zone difficult. In the western Himalaya, the Indian government has built a number of military roads in Ladakh and in the adjoining plateau regions of Spiti and Lahaul. Many of these roads are now open to civilian traffic, and their presence has opened the plateau to regional transportation. Elsewhere in the Himalaya, though, the plateau areas are generally devoid of vehicular traffic. Several passes, called *la*, enter the zone from the south, crossing the Great Hima-

Indus River tributary, Zanzkar.

Kathmandu, Nepal.

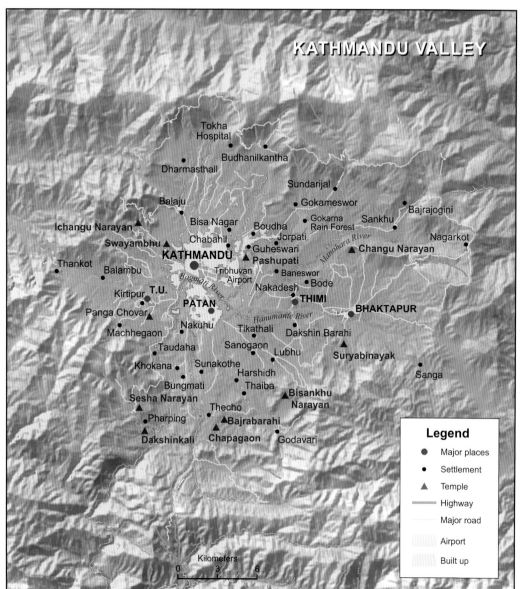

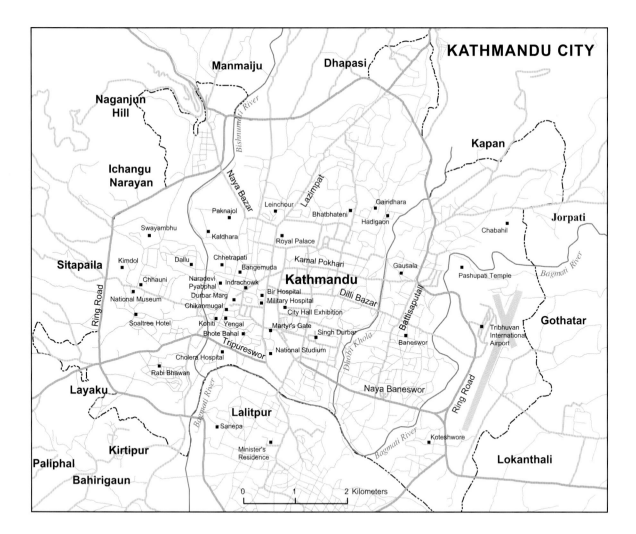

(pages 20–22) MAJOR VALLEYS AND CITIES

*Kashmir Valley* in the western Himalaya, and *Kathmandu Valley* and *Pokhara Valley* in Nepal, are tectonic depressions that formed ancient lake beds. In Kashmir, several small remnant lakes, including Dal Lake, Nageen Lake, and Wular Lake, are distinctive features of the present-day valley floor. The Kathmandu Valley is drained by the Chobar Gorge. Both valleys also contain sizable settlements, including two of the largest cities in the Himalaya—*Srinagar* (estimated population 710,000, 1991 census) and *Kathmandu* (estimated population 1,093,414, 2001 census). Kathmandu is growing at an exceedingly fast rate (4.8 percent per annum) due to in-migration from the hills. Kathmandu Valley also contains two traditional city-states: *Patan* and *Bhaktapur*. *Pokhara Valley* in central Nepal contains lake Phewa Tal, the town of *Pokhara,* and rich agricultural land.

Dal Lake, Kashmir Valley.

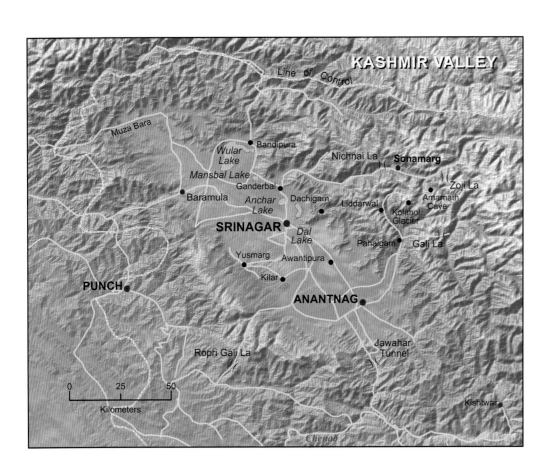

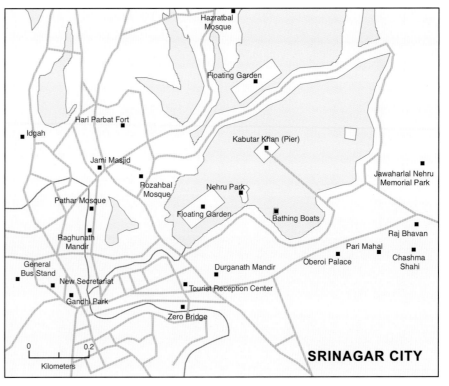

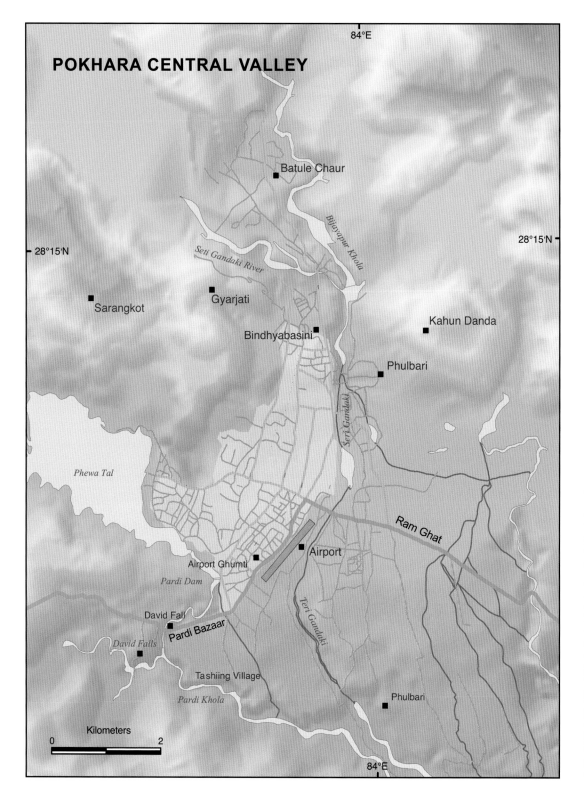
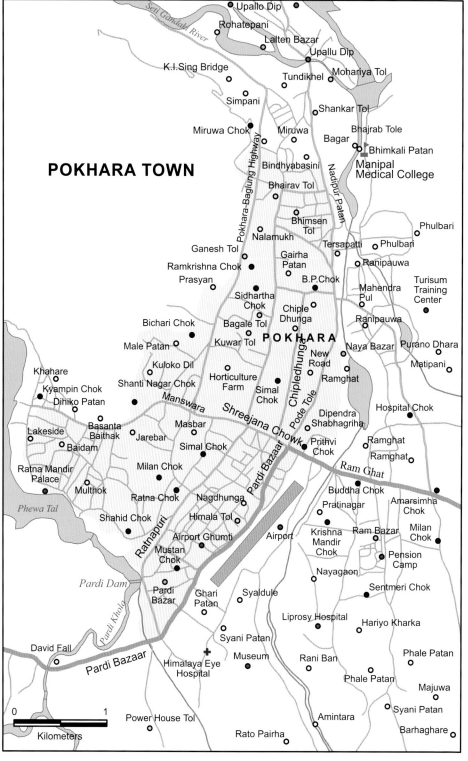

laya at heights ranging from 4,000 to 6,000 meters. The major passes include the 4,373-meter Jelep La in the Chumbi Valley of Sikkim; the 5,150-meter Laitsawa Pass in Bhutan; numerous important routes in Nepal in the areas of Dolpo, Khumbu (Mount Everest region), and Mustang; and the crossings of the Zanzkar and Ladakh ranges in India at 4,650-meter Baracha La and 4,276-meter Kun Zum La.

South of the plateau zone, extending west to east across the entire range, lie the magnificent summits of the Great Himalaya. This highland zone composes an almost contiguous rampart of ice, snow, and rock and, in the minds of most people, conjures the strongest images of the Himalaya. Here, ancient Tethys Sea deposits more than 15 kilometers thick have been uplifted to expose the crystalline core of the mountains. Amid the high peaks are deep gorges, secluded tributary valleys, and saddle passes, providing difficult avenues for travel and remote places for human settlement. The altitude, climate, and harsh terrain preclude many people from living in this zone. In the

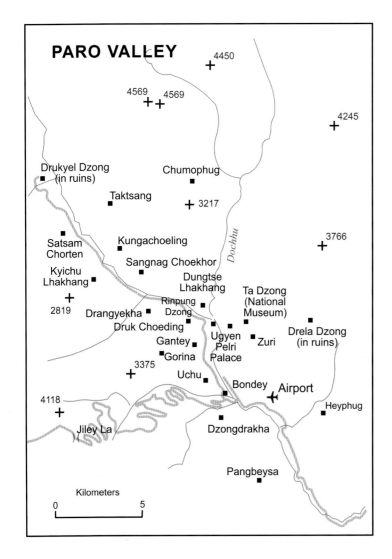

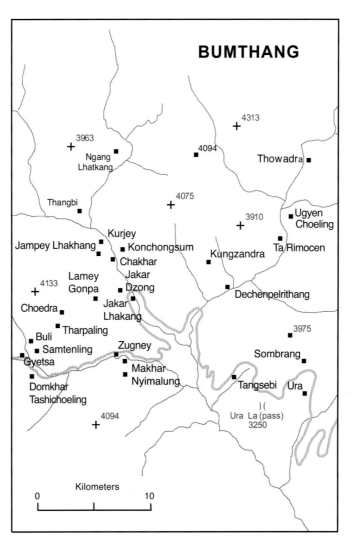

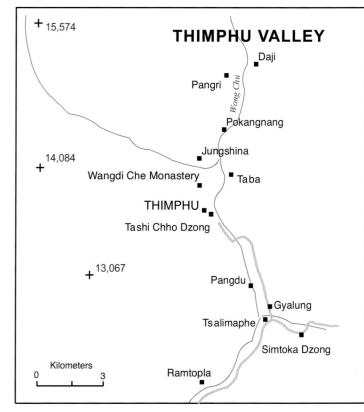

BHUTAN VALLEY SETTLEMENTS: PARO, BUMTHANG, AND THIMPHU

Overall, Bhutan has lower population densities than those found in the western and central sectors of the Himalaya. However, the broad, fertile valleys of the middle mountains zone are densely populated, reflecting their agricultural potential as well as their historical political importance. The presence of *Dzongs*, or fortress settlements, signifies the historical feudalism of Bhutanese society and the overwhelming presence of monastic communities.

Paro Valley, Bhutan.

Zanzkar and Spiti valleys of India and in the upper Kali Gandaki Valley of Nepal, where clear skies prevail for much of the year, the abundant sunshine makes agriculture possible where water for irrigation is available. In these places, small villages, constructed of mud and stone houses, are clustered around old Buddhist monasteries. During the long winters, snow in the high passes isolates many of the valley settlements and makes travel difficult or impossible. For example, to leave Zanzkar in the winter, one must walk for a week or more across the dangerous frozen surface of the Zanzkar River, moving from the

Bumthang Valley, Bhutan.

TOWNS OF THE HIMALAYA: GANGTOK, SIMLA, AND DEHRA DUN

The rise of urban centers is a relatively recent phenomenon in the Himalaya, where most of the population remains rural and agrarian. Some Himalayan towns, such as Simla and Darjeeling, originated with the British, who developed the settlements as summer retreats during the colonial period. English officers and their families, as well as other military, clerical, and business people, fled to the higher altitudes of the Himalaya during the hot months. Simla became prominent as the summer capital of the British Raj. Other Himalayan towns have a more recent origin, their growth stemming from migration from the countryside to the city. In Nepal, for example, where rural–urban migration rates are high, the urban population increased from 3.6 percent in 1961 to 12.7 percent in 2001. Wherever roads are built in the Himalaya, towns spring up, so there is a close correlation between accessibility and development.

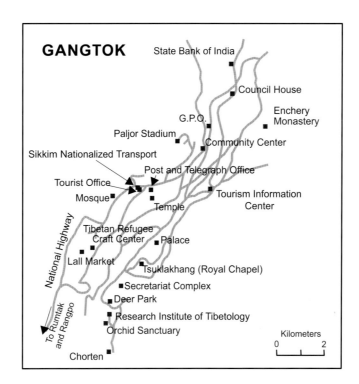

Pedestrian mall, Simla.

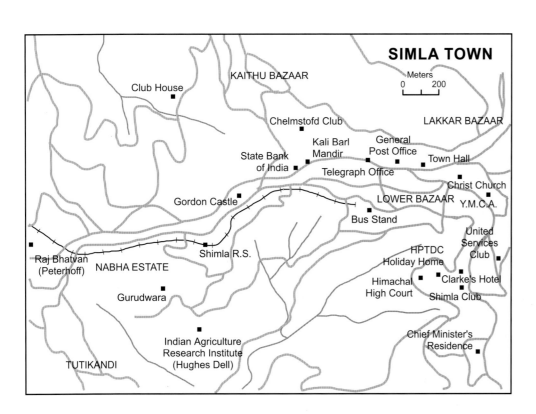

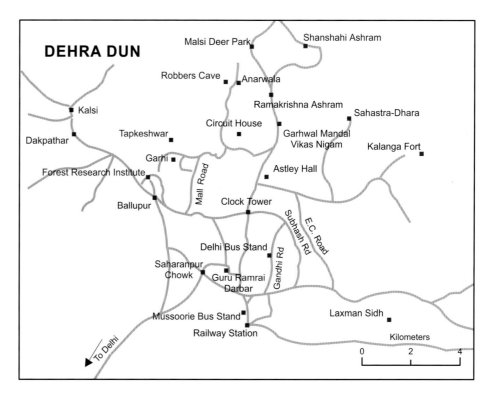

## TARAI TOWNS OF NEPAL: BIRATNAGAR AND NEPALGANJ

The lowlands of Nepal near the Indian border are one of the fastest-growing regions in the entire Himalaya. Border towns such as Biratnagar and Nepalganj play an important role as transit routes and trade depots and have been settled in recent decades by large numbers of Indians as well as Nepalese migrants from the hills. The border between India and Nepal is effectively open for both nationalities, allowing free movement and trade between the two countries. This has led to a rapid increase in the populations of tarai towns (2001 population estimates: Biratnagar, 222,279; Nepalganj, 83,111). Population growth in the tarai towns, as in the hill towns, leads to serious management and planning issues related to poverty alleviation, infrastructure, and urban services.

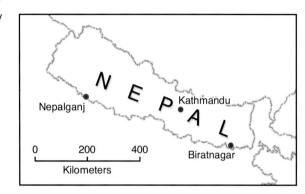

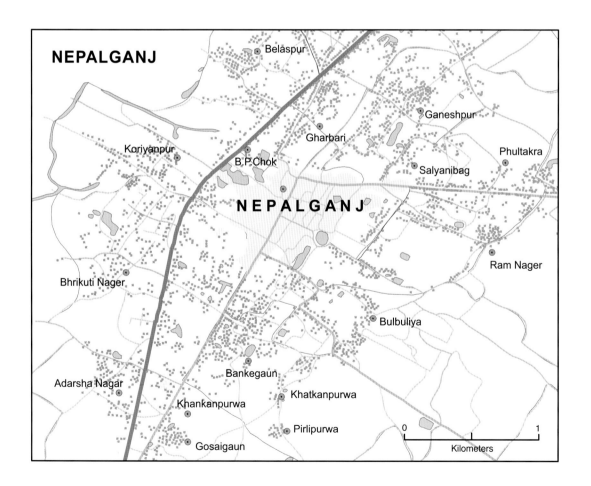

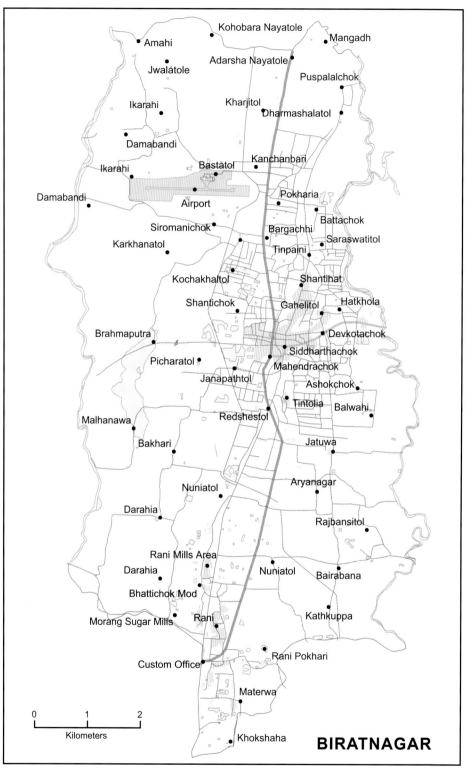

interior of the Zanzkar Valley to the road head near Chiling in Ladakh, along a treacherous canyon trade corridor that is many centuries old. In Nepal and Bhutan, the Great Himalaya zone is used primarily as a summer grazing area by the seminomadic peoples who live for most of the year in villages at lower altitudes. The upper flanks of such great massifs as Dhaulagiri, Annapurna, Everest, and Kanchendzonga host significant populations of seasonal herders—the Magar, Gurung, and Sherpa tribes, for example—but provide few resources for permanent villages. In recent years, the Great Himalaya zone has become a popular destination for mountaineers and trekkers, and a number of national parks have been newly established there.

The bulk of the human population in the Himalaya resides in the intermediate middle mountains zone, which falls south away from the Great Himalaya in terrain ranging from 1,500 to 4,500 meters in elevation. The middle mountains landscape is heavily dissected by rivers flowing from the higher mountains and contains numerous ridges and valleys in a complex and rugged topography. Several prominent and discrete ranges occur in this zone, including the Pir Panjal in the northwestern Indian Himalaya, the Mahabharat Lekh in Nepal, and the Black Mountains of Bhutan. These ridges, by virtue of their greater height, stand above much of the rest of the middle mountains, but the entire zone is mountainous and in full view of the high peaks of the Great Himalaya. Numerous rivers fed by the melting snow and glaciers converge in the middle mountains zone to form the Himalaya's great river systems: the Indus-Sutlej in the westernmost region; the Alaknanda-Bhagirathi in the Garhwal region; Nepal's Karnali, Narayani, Gandaki, Kosi, and Arun rivers; the Tistha in Sikkim; the Amo-Sankosh and Manas in Bhutan; and the Brahmaputra in Arunachal Pradesh, which forms the eastern boundary of the Himalaya. These river systems play a vital role in shaping the topography of the middle mountains zone, as well as providing water for irrigation and potential energy for hydropower development.

One of the most striking aspects of the middle mountains landscape is how intensively it is managed by human society

River crossing, central Himalaya.

The Regional Setting 27

—in some places, to the point of exhaustion. Farm communities dot the ridges and spread across the lower slopes, interspersed by areas of forest and cultivation. The forests are most intact where population densities are low, but even in heavily settled areas, forests are present as conservation areas, as religious sanctuaries, and as village common lands where fodder is collected and livestock is taken to graze. Cultivated lands are widespread in the river valleys, where irrigation water is available to grow rice, and alluvial terraces provide fertile soils. The mountain slopes have been carved over centuries into cascades of level terraces, so that in places, more than 2 kilometers of vertical mountain is covered in carefully managed farm fields. Considering how difficult it is to hoe a small garden by hand, imagine the labor required to shape an entire Himalayan mountainside into a series of flat surfaces. The extensive terraces are a compelling sight, but they highlight the growing need for agricultural land to feed an ever-expanding human population. Population growth rates of 2 to 3 percent a year are common in the Himalaya, and in some areas, such as the outer foothills, annual growth rates exceed 4 percent. The farm terraces also reflect a remarkable knowledge about the land and soil and display the advanced engineering skills of the traditional agricultural societies.

The middle mountains zone descends in altitude until it forms a line of outer foothills, known in India as the Churia Hills, in Nepal as the Siwaliks, and in Bhutan as the Duars (meaning "gateway"). The foothills, in turn, give way to a southward-sloping piedmont plain (in Nepal, known as the *tarai*), which forms the northern extension of the Ganges-Brahmaputra Plain. A system of tectonic depressions, called *dun* valleys, occurs in the foothills; these are filled with alluvium deposited by the rivers and make good farmland (for example, the Dang Valley in Nepal). Significant forest areas remain in the foothills zone, mainly because this area has avoided intensive human settlement, but these forests have

Potato market, Kulu Valley.

been threatened by intensive logging operations and a steady flow of migrants from the mountains since the 1960s. The outer foothills and plain are viewed as resource and settlement frontiers by many of the Himalayan countries, and agricultural, industrial, and urban expansion is occurring there at a rapid rate. The many new towns and roads that have sprung up in this zone provide convenient access to once-remote regions in the high mountains and to the bustling plains of northern India. Gateway towns are developing quickly all across the southern perimeter of the Himalaya, lending an urban and industrial look to the lowland landscape.

There is much interest in the current state of the Himalayan environment, with diverse scenarios proposed about the level of land degradation and the reasons for it. Amid this confusion, it is clear that the extraordinary natural and cultural diversity in the Himalaya does not permit any single explanation for the ecological conditions across the entire range. Many places are subject to almost catastrophic levels of soil and water degradation, declining farm productivity, and increasing human impoverishment. These trends have been in place for a number of decades, and their resolution remains distant. Elsewhere, careful husbandry of natural resources is practiced, local economies are progressive and sustainable, and a productive landscape remains largely intact. In general, though, the combination of geologic instability, expanding human populations, and accelerating resource extraction by national development takes its toll and results in greater vulnerability of the natural and human ecosystems. Notably, the most serious land degradation occurs in places where local communities have lost the authority to manage their own resource environment, where people are so impoverished that their livelihoods become acts of desperation, and where infrastructures supporting the industrial economy are the most exploitative. Under such circumstances, it is difficult to imagine a sustainable future where the needs of both society and nature can be met. In light of the pressing environmental trends, new conservation initiatives are proposed that range from grassroots economic strategies to large national parks. These offer considerable hope amid the challenges of a new millennium.

A common feature of most successful Himalayan conservation programs is the awareness that environmental preservation must be tied to appropriate economic and social opportunities, whereby people can manage on a sustainable basis their most basic needs as well as their cultural aspirations. In this vein, maintaining cultural diversity is absolutely necessary for managing biological diversity. Many people recognize that the Himalayan environment is unique, its size and diversity are majestic, yet it is the rich assemblage of cultures that transforms the wild and scenic beauty of the mountain landscape into a place of human dimension and devotion and ensures its continued presence in the face of inevitable change. Nature and society together in the Himalaya compose an elegant and challenging mountain landscape and shape one of the most stunning places on earth.

# PART TWO  The Natural Environment

# Geology

## GEOLOGIC HISTORY

The Himalaya is one of the youngest mountain ranges on earth. Its geologic history began 60 million year ago, when India first collided with Asia, but most of its altitude was gained during the last 2 million years. The mountains continue to grow today amid widespread and frequent earthquakes. These seismic tremors signify the geologic forces that produced such spectacular highland terrain, but they also make the Himalaya a tectonically dangerous place to live. The mountains' propensity for seismic disturbances, their steep terrain, gravity, and the forceful movement of water across the rugged slopes combine to create what geologists call a "high-energy" environment. The kinetic potential contained within the contours of the Himalaya is enormous. The summits and gorges, the long lines of undulating ridges, and the diverse terrain that we see as the actual mountains are merely the outer skin of a geologic plate that underlies the region and is up to 75 kilometers thick. The movement of this crustal fragment during the past 60 million years, with the South Asia plate submerging beneath that of the Asian continent and lifting the oceanic crust of the ancient Tethys Sea along the way, is responsible for the formation of the mountains. They continue to grow because the Indian plate maintains its northward drift into Eurasia at a speed of about 2 centimeters per year.

Our knowledge of the physical development of the Himalaya is tied to the theory of continental drift, which explains the world's landforms by the breakup of primordial supercontinents and movements of the earth's lithosphere. In the case of the Himalaya, the breakup of Gondwanaland over the past 500 million years contributed fragments of crust that slowly drifted northward toward the Siberian shield, eventually colliding with the Asian continent and causing massive upthrusts along the advancing edge of the Indian plate. The northward movement of India toward Eurasia began about 130 million years ago, contracting the intervening Tethys Sea, and the actual collision of India and Asia began some 60 million years ago during the early Tertiary period. As a result of this collision, the oceanic crustal rocks and sediments of the Tethys Sea were thrust upward along an interface known as the Indus-Yarlung suture zone. The present-day Indus and Yarlung-Tsangpo (Brahmaputra) rivers follow this alignment and wrap the tectonic Himalaya in a geologic embrace that stretches west to east for 2,700 kilometers. The fact that these two rivers demarcate the tectonic rendering of the Himalaya provides a useful reference for the geographic boundaries of the mountains. To the west of the Indus River, at the juncture of the Nanga Parbat uplift, is the Shyok suture zone, named after an important tributary of the Indus. Geologically, this

⌃ (part opener) A tributary of the Sutlej River flows through a narrow gorge in the western Himalaya.

« THE HIMALAYA AND TIBETAN PLATEAU FROM THE SOUTHEAST

A digital terrain model created from elevation data and NASA space imagery overlooks the length of the Himalaya from a point above Myanmar looking northwest. (The Hindu Kush Range in Afghanistan is situated at the top left of the image, and the Tarim Basin and Takla Makhan Desert in China's Xinjiang Province is at the upper right.)

The main thrust zone of the central Himalaya contains several deep valleys hemmed in by high peaks.

The Natural Environment 33

DRIFTING CONTINENTS AND FORMATION OF THE HIMALAYA

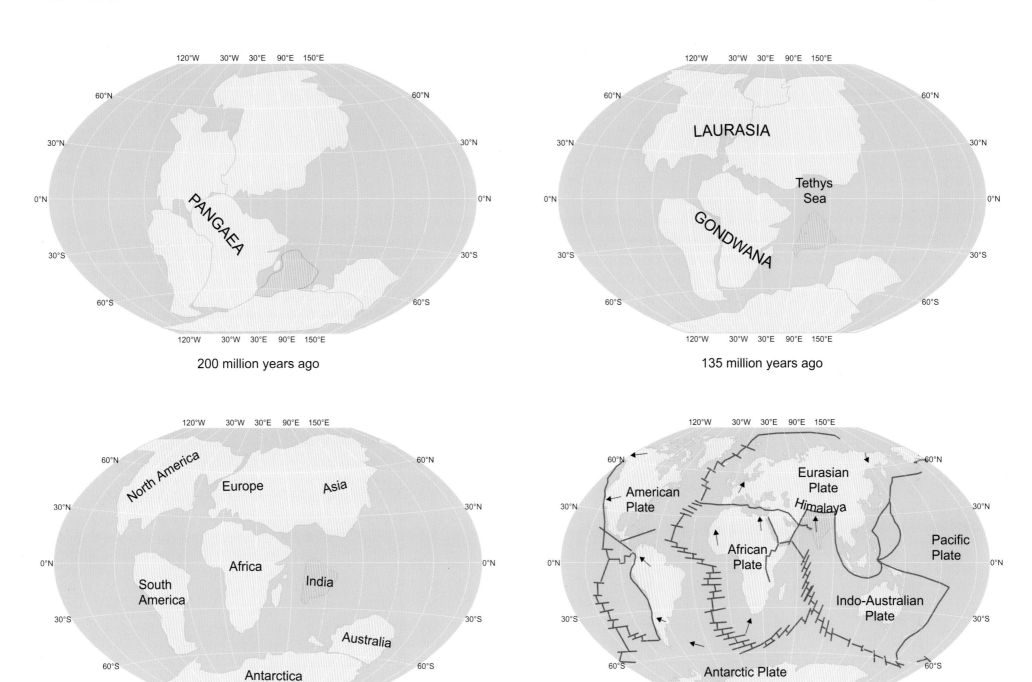

zone separates the Himalaya from the Karakoram mountains, although the two ranges are commonly joined in the rendering of a Pan–South Asian highland system.

When the Karakoram is included in the Himalayan chain of mountains, the system includes all fourteen of earth's peaks over 8,000 meters and hundreds of others greater than 7,000 meters in elevation. Nine of the 8,000-meter peaks are in Nepal, including the world's highest mountain, Mount Everest (8,850 meters). The world's second-highest peak, located in the Karakoram Range in Pakistan, is K2 (8,611 meters). Nepal contains all or part of eight other 8,000-meter peaks: Kanchendzonga (8,598 meters), Lhotse (8,501 meters), Makalu (8,470 meters), Dhaulagiri (8,172 meters), Cho Oyu (8,153 meters), Manaslu (8,125 meters), Annapurna (8,091 meters), and Kao-seng-tsan Feng (8,013 meters). The remaining four peaks above 8,000 meters are in the Karakoram: Nanga Parbat (8,125 meters),

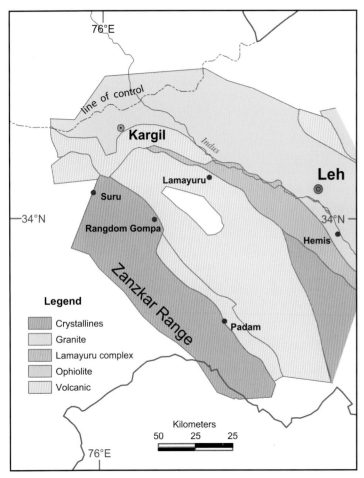

GEOLOGY OF ZANZKAR AND INDUS VALLEY

The Indus River marks an important suture zone defining the geologic boundary of the Himalaya. The western regions of Zanzkar and Ladakh straddle the Indus Valley region and are marked by the juncture of the high, dry trans-Himalayan plateau of Ladakh and the Great Himalaya of Zanzkar, with its icy peaks and high valleys. The geology of this region reflects its position in the crystalline thrust belt.

» (overleaf) SATELLITE MAP OF THE HIMALAYA

Land cover data from satellite imagery overlays NASA elevation data to show landscape features across the range.

Vertical walls of the central upthrust zone.

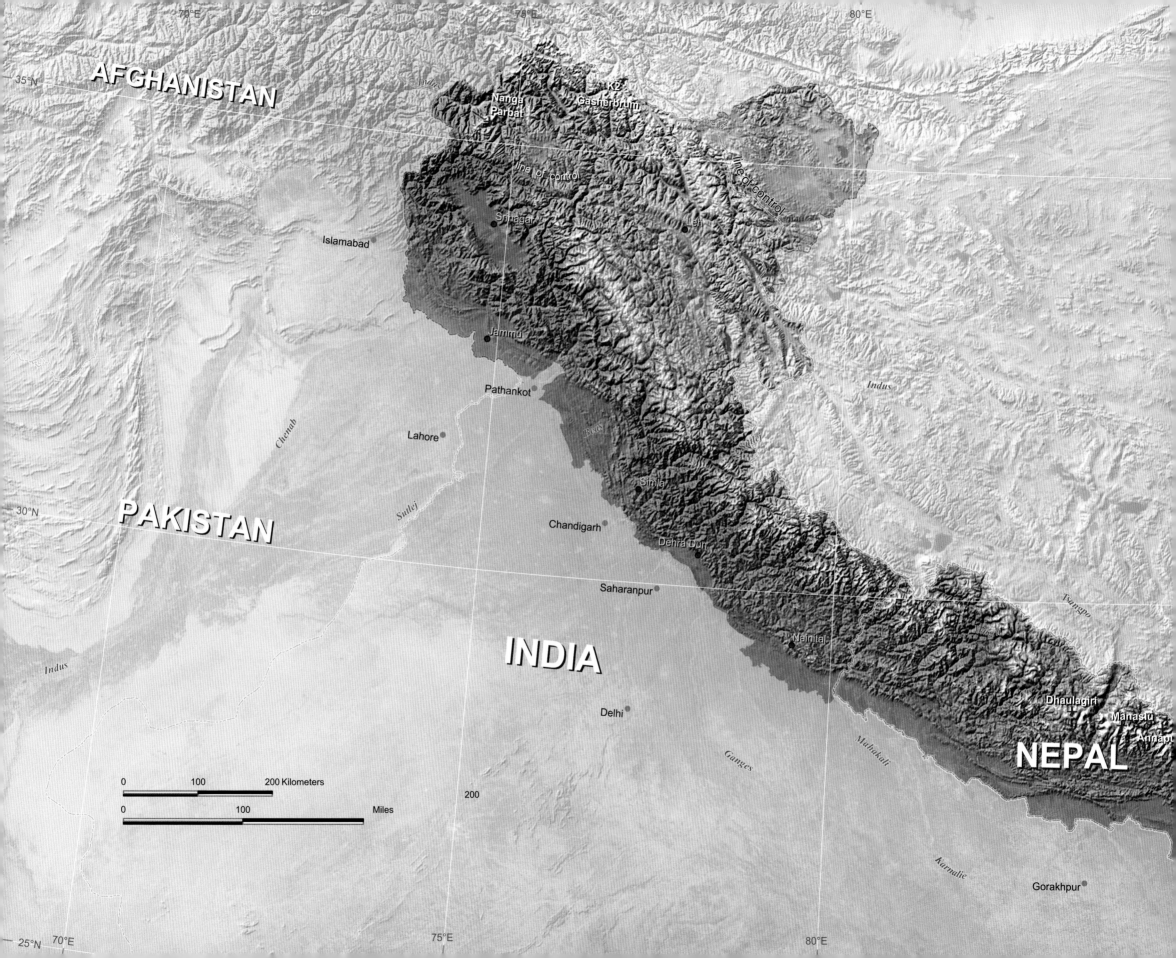

Hidden Peak (8,068 meters), Gasherbrum (8,060 meters), and Broad Peak (8,047 meters). These prominent mountains are recognizable massifs in the Himalayan skyline and represent the outermost extrusions of the huge crystalline masses that compose the High Himalaya.

The tectonic Himalaya extends in a northwest to southeast direction from the Indus River in present-day Pakistan, near the Nanga Parbat summit, to the Brahmaputra River in India's northeastern mountains, near the Namche Barwa summit. West and east of these summits, the Himalaya joins with the other lineaments of the circum-Indian mountains to create the highest places on earth. It is likely that the Indian plate first struck Eurasia in the northwestern part of the Himalaya, where the mountains bend southward around Nanga Parbat; 20 million years later it struck in the eastern sector near Namche Barwa, where the Himalaya joins with the highlands of northern Myanmar. This would help account for the directional strike of the range, as well as for the distribution of different rock types and ages and tectonic structures.

The remote areas of this highland region have been known

Exposed strata, Zanzkar.

for millennia by its native peoples, who in the course of their residence have come to ascribe supernatural powers to its compelling geologic characteristics. In the early 1800s, colonial explorers working for the East India Company began preliminary mapping of the Himalayan frontier for the British Empire. In 1851, with the establishment of the Geological Survey of India in Calcutta, a series of systematic geologic studies was undertaken. In 1907, a comprehensive geologic map of the Himalaya was created by Burrard and Hayden and published by the Government of India Press. These efforts laid the foundation for more recent surveys of the Himalaya, including those of Swiss geologists Augusto Gansser and Toni Hagen, both of whom contributed comprehensive overviews of the Himalaya that remain benchmark studies of the region's geology.

The range of relief from high to low elevations in the Himalaya is unsurpassed by other mountains, but the geologic formation of its structures is still not fully understood. The earliest geologic inquiries relied on the dating of exposed rock layers based on fossil findings. More recent investigations consider the underlying structure of the Himalaya's main thrust sheets to explain the geologic divisions of the range. They distinguish four main zones, from north to south: the thick-crusted Tibetan or Tethyan zone, sometimes called the trans-Himalaya, which constitutes a small part of the Himalaya proper but is an important outer margin of the extensive Tibetan Plateau; the Great or High Himalaya, which caps the complicated geology of the main thrust zone in a series of lofty peaks and snowy summits; the Lower or Lesser Himalaya, or the middle mountains zone, which occupies a 65-kilometer band of intermediate hills striking across the midsection of the range; and the Outer Himalaya or Siwalik foothills zone, which is made up of a series of low-altitude ridges separated by alluvium-filled tectonic basins.

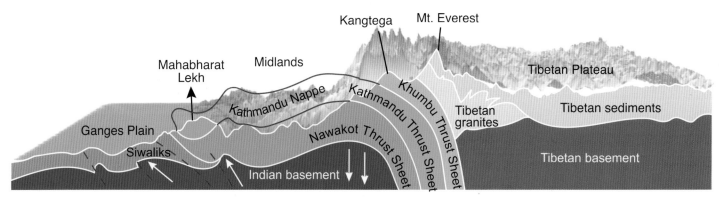

GEOLOGIC CROSS SECTION: 87 DEGREES EAST LONGITUDE (MOUNT EVEREST)

The main boundary fault north of the Dun Valley is located approximately where the middle mountain zone meets the outer foothills. This is one of the most tectonically active zones in the Himalaya.

The Natural Environment 39

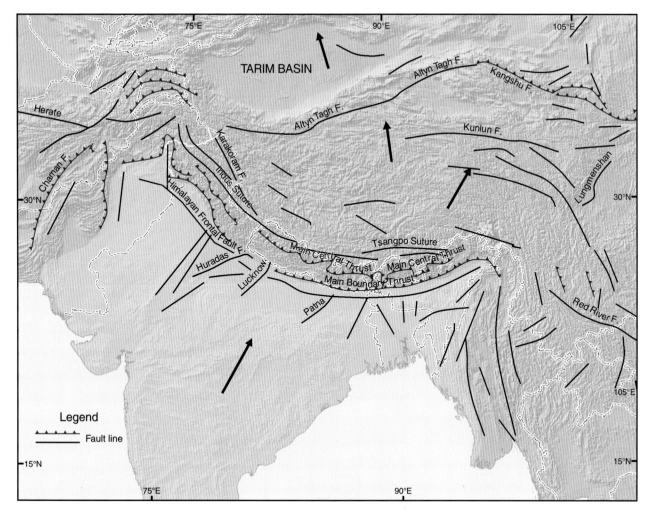

⌃ NORTHWARD DRIFT OF INDIA

The collision of India and Asia, which began about 60 million years ago, results from the northward drift of the subcontinent at a current rate of 2 centimeters per year. This movement formed the Himalaya.

⌃ Wooden houses in the village of Chitkul cling to the uplifted strata of the Sangla Valley, Indian Himalaya.

« Granite and schist peaks of Garhwal Himalaya.

The complex geology of the Lower and High Himalaya is partially attributed to the great rock sheets, called nappes, that have been displaced many kilometers. The nappes essentially represent scrapings of the Indian continent that have been forced backward (south) atop the forward (north)-moving plate as India passes beneath Asia. The mountains are crossed by rivers that cut through the strike of the range, producing some of the deepest gorges in the world. The rivers that cut the great transverse valleys predate the uplift of the Himalaya and originate in the Tibetan zone; thus, they have been continually eroding the mountains even while they have grown. The main rivers breaching the Himalaya include the Indus (1,200 meters above sea level and 22 kilometers from Nanga Parbat), the Kali Gandaki (1,500 meters above sea level and 7 kilometers from Dhaulagiri), and the Trishuli (1,800 meters above sea level and 13 kilometers from 7,245-meter Langtang-Lirung Himal). These rivers, in turn, are fed by melting glaciers. Some of the world's largest glaciers are located in the South Asia

### GENERALIZED GEOLOGY OF THE HIMALAYA

A simplified overview of the complex geology of the Himalaya reveals a regional structure of tectonic zones from north to south: the Indus suture line, the Tibetan or Tethyan Himalaya, the Great or High Himalaya, the Lower Himalaya, and the Outer Himalaya or Siwalik foothills. These zones result from the tectonic processes of uplift, are distinguished by contrasting rock layers and mineral content, and are separated by faults and thrust boundaries, including the main central thrust, separating the trans-Himalaya and Tibetan zone from the High Himalaya, and the main boundary fault, separating the Lower Himalaya and the Outer Siwalik zones. The prominent rocks of the Himalaya originated in the sediments of the ancient Tethys Sea and the crystalline structures of the central upthrust zone, where tremendous heat and pressure metamorphosed the rock layers. The distribution of rock types shown on the map reflects the underlying tectonic structures.

**Legend**
- Trans-Himalayan granite
- Oceanic crust ophiolites
- Anatectic granites
- Tethyan sea sediments
- Lower Himalaya sediments
- Siwalik sediments and alluvium

mountains: Siachen Glacier (72 kilometers), Hispar Glacier (61 kilometers), and Baltoro Glacier (58 kilometers), all located in the Karakoram.

Geologic studies show the Himalaya to be composed of an exceedingly complex landscape of diverse origin and structure, struck through with faults and fissures and hotspots, twisted by folds and thrust belts, and overfilled in places with sediments and surface deformations. These local irregularities, which occur on a mammoth scale in the Himalaya, conform to a fairly consistent overall geologic structure. They contribute to the mountains' overwhelming size as well as to their compelling beauty and rich mineral resources. The age of the Himalaya generally diminishes from north to south. The oldest geologic structures and rock materials occur along the axis of the main crystalline thrust sheet, which is located in the High Himalaya south of the Indus-Yarlung suture, often underlying the sedimentary deposits of the northernmost Tibetan or Tethyan Himalaya. Where the high peaks join the uplifted Tibetan Plateau, the Himalaya is covered by sedimentary rocks that originated in the uplift, compression, and erosion of the Tethys seabed. The tectonic boundary between India and Asia, the so-called Indus suture line, is characterized by a mix of sedimentary, metamorphic, and volcanic rocks. The adjoining Tibetan Himalaya is about 15 kilometers thick and contains fossil-laden marine rocks that once constituted the northern margin of the Indian continent.

South of the Tibetan Himalaya, in the High Himalayan zone, are located the world's highest mountain peaks. They tower above the geologic substrata of this zone as skyward extensions of a crystalline sheet more than 10 kilometers thick and 100 kilometers from north to south. Stretched over much of this inner-core crystalline mass, like a thick skin, is a layer of sedimentary rock that has its origins in the Tethys seabed. In the north, where the crystalline structures meet the Tibetan Himalaya, is a zone of mica and tourmaline outcroppings—a product of the melting of the Indian continental crust. Geologists puzzle over many of the local deformations in the High Himalaya, but overall, the zone provides a relatively simple tectonic picture. A huge crystalline thrust sheet extends for much of the length of the Himalaya Range, supporting the sedimentary surface landforms of the high mountains; it is in contact with the Tethyan Himalaya in the north and overthrusts the Lower Himalaya in the south.

The crystalline upthrust of the High Himalayan zone is thought to have coincided with the huge horizontal compression of the northern edge of India as it drifted northward and made contact with Asia. This compressed rock, known as the "root zone" of the Himalaya, was actually uplifted as a great nappe sheet and moved southward over the intervening materials, so that geologically, the High Himalaya looms over the Lower Himalayan zone. The southern reaches of the original nappe sheets eroded as India continued its northward drift, exposing the Lesser Himalaya in the foreground and leaving behind the crystalline substructure of the High Himalayan peaks. The principal upthrust occurred quickly in geologic reckoning, about 600,000 to 1 million years ago, and this phase marks the period when the present-day landscape of the High Himalaya took most of its shape. If we were to compare geologic time to a year of human time, the upthrust that formed the Himalaya took about an hour and a half. The scale of this upheaval is nonetheless monumental, with geographic implications for all of Asia. The high peaks of the main thrust zone form a huge watershed divide between the Gangetic and Indus plains to the south and the northern plateau of Tibet, which has risen by about 4,500 meters in the last million years and today is the highest landmass on the planet. This divide also has climatic and cultural significance.

Many of the rocks found in the High Himalaya, including the precious gemstones associated with the crystalline structures, have their origin in the heat and pressure created by this upheaval. In addition to quartzite rocks, the zone contains immense deposits of granites, including those with a high mineral content and value (for example, the leucogranites, which include tourmaline), micas, marbles, and gneisses. Young granite intrusions are especially prominent in this zone. They often form the highest peaks, their flanks embedded with stria-

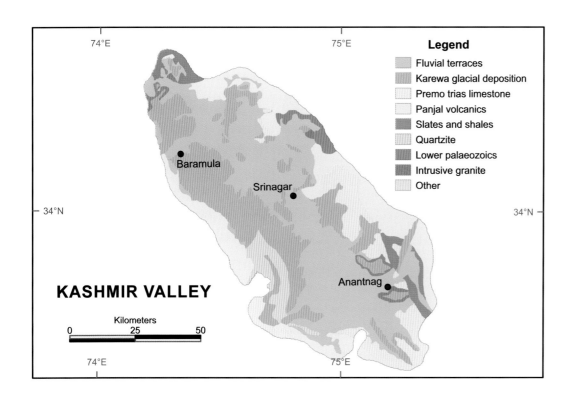

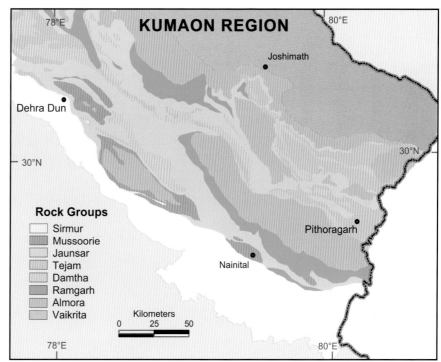

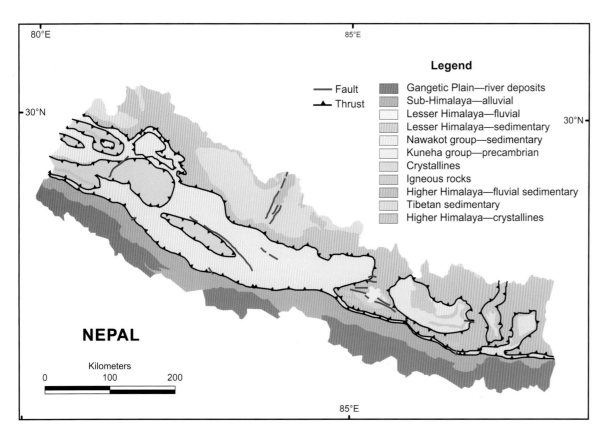

## HIMALAYAN GEOLOGY

Broadly speaking, the regional geology of the Himalaya is explained by major thrusting and folding of rock sheets, with the crystalline rocks of the High Himalaya being transported south along the main central thrust zone. In the northern Tibetan zone are primarily sedimentary rocks, and the southern foothills are formed from conglomerates and sediments deposited by rivers. A large number of faults are observed on satellite images and indicate that the most active zones are in the Lower Himalaya and Siwalik foothills.

Agricultural terraces cascade down steep mountain slopes in eastern Nepal.

tions of contrasting color that denote the extrusion of minerals through geologic fissures. The vertical uplift continues in the Himalaya at a rate ten times faster than that of the European Alps, and its exposed strata bear testimony to its ancient low-altitude origins: flora and fauna fossils found at elevations above 6,000 meters indicate a tropical climate.

During the latest phase of mountain building, about 20,000 years ago, the Himalaya experienced an ice age. The glaciers gained size in the high elevations and scoured the mountain slopes into the dramatic ridges, cirques, and hanging valleys we see today. Numerous lakes formed from the glacial melting that followed the last ice age, many of which have since dried in the desertlike conditions north of the main peaks. In Tibet, the evaporating lakes left behind huge salt pans, which the Tibetans traditionally mine for export to India along the great Himalayan salt route. The depositional hills of glacial till, called moraines, form dams in the High Himalaya and create lakes from the melting glaciers. These natural walls of glacial debris occasionally burst, allowing the impounded water to escape in cataclysmic floods that threaten life and land downstream at lower altitudes. These glacial lake outburst floods constitute a significant hazard in the high mountains.

Located south of the High Himalayan zone, between the main central thrust in the north and the main boundary fault in the south, is the Lower or Lesser Himalaya. This region is commonly referred to as the middle mountains, suggesting its intermediate altitude. Much of the geology of the zone is composed of compressed shield material from India and sediments from the Tethys Sea. In keeping with the general chronology of Himalayan geology, which shows the age of thrusting to diminish from north to south, the Lower Himalaya is of more recent formation than the High Himalaya. The main boundary fault marking the southern boundary of the Lower Himalaya corresponds to where the thrust sheets of the High Himalaya made contact with the outermost Himalaya. Two important ranges appear in this contact zone—the Mahabharat Lekh in the middle mountains and the Siwalik in the Outer Himalayan foreland. Independent thrust sheets also formed in some places

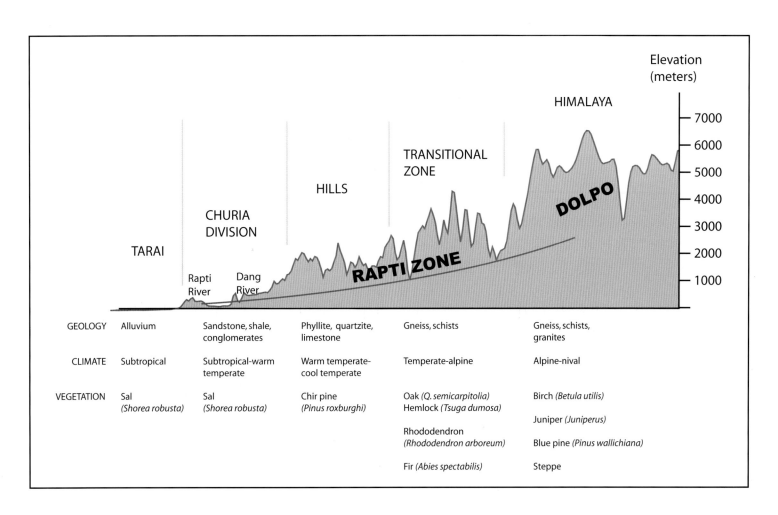

GEOECOLOGY OF THE WESTERN MOUNTAINS

A transect of western Nepal (82 degrees, 20 minutes east longitude—Rapti zone to Dolpo) shows the correspondence of geology, elevation, climate, and vegetation. The distribution of life zones by altitude is a premier characteristic of the Himalayan environment.

in the Lower Himalaya, and these too overthrust the outermost sub-Himalaya, creating a zone of high tectonic activity.

There is a great deal of variation in the geology of the Lower Himalayan zone, but in many places the inner thrust sheets dip steeply to the north, creating formidable natural barriers in the terrain. With the exception of the Mahabharat Lekh, which consists of hard rocks and steep slopes, the Lower Himalaya generally contains rocks that are less resistant than those found in the High Himalaya; hence the hillsides tend to erode more easily and to be more gradual, with deeper soils. The Mahabharat Lekh, however, composes such a formidable rock barrier that many rivers flowing south from the High Himalaya are forced west or east before breaching the range and continuing south. The entire Lower Himalayan zone continues to be uplifted at a geologically fast rate; the Kathmandu Valley in Nepal, for example, has risen over 200 meters in the past 200,000 years.

South of the main boundary fault is the so-called Outer Himalaya or Siwalik foothills zone, which is associated with some of the most active tectonics in the region. This zone consists of sediments derived from the uplift and erosion of the main Himalaya. It straddles the Himalayan frontal thrust, which marks the outermost boundary of the Himalaya, and overlooks the alluvial tracts of the Gangetic Plain in India. Much of the Outer Himalayan zone is occupied by the Siwalik Range, which is a series of foothills formed from faulting and folding, interspersed with sediment-filled tectonic basins. These basins, such as Dang Valley in Nepal and Dehra Dun in India, are called dun valleys and constitute agriculturally rich alluvial deposits. Whereas the southern edge of the Outer Himalaya gradually declines in altitude to meet the Indian plains, the northern boundary of the zone is distinguished by the 25-kilometer-wide main boundary fault, which is a steep north-dipping fault and thrust zone. Much like the dendritic tributary patterns of a watershed, the main boundary fault zone constitutes a hierarchy of subsidiary faults and thrust branches. These diverge in the west and converge onto the main fault in the east, exposing the tectonic origins of the Outer Himalaya. The tectonic movements of the Siwalik zone are especially active, and the entire zone is subject to numerous tremors and earthquakes.

## GEOLOGIC REGIONS

### Western Sector

In the western section of the Himalaya, situated between the Indus and Sutlej rivers, is a 500-kilometer stretch of mountains that composes much of the Indian states of Kashmir and Himachal Pradesh, as well as a tiny portion of Pakistan. The western sector is dominated by Nanga Parbat, which, at 8,125 meters, is the highest peak of the Kashmir Himalaya and the ninth highest peak in the world. Nanga Parbat anchors the western Himalaya both geologically and visually, and it is the northernmost outcrop of the Indian continental crust. Many geologic investigations, mainly by German scientists, have centered on Nanga Parbat, in part because it is a very young mountain that prominently displays the quintessential traits of mountains in its exposed strata. Perhaps most importantly, though, Nanga Parbat is the pivot point for the western Himalayan syntaxis, wherein the mountains bend from northeast to southwest along the Indus suture. It thus occupies a critical position in the tectonic structure of the entire Himalaya Range.

The base of Nanga Parbat consists of metamorphosed gneisses, with compositions that include muscovite, garnet, cordierite, and other leucogranites. The most recent phase of the Nanga Parbat uplift produced tourmaline-bearing granites. The gemstones of Nanga Parbat are of commercial interest, but the massif is best known for its dramatic and isolated silhouette amid the terrain of the western Himalaya. The closest 8,000-meter Himalayan peak in an easterly direction is Dhaulagiri in Nepal, a distance of about 1,100 kilometers. To the north is the Indus Valley, which, at 1,300 meters above sea level, provides a striking topographic contrast to the altitudes of Nanga Parbat. The mountain is bounded by the Ladakh and Kohistan formations, which constitute much of the western

Himalayan region, and it appears conspicuously at the juncture of the High Himalaya, the Kohistan-Ladakh mountain arc, and the Karakoram, a circumstance that is tectonically accounted for by the Indus-Tsangpo suture and the Shyok suture zones located nearby.

South of Ladakh and in direct contact with Nanga Parbat is the crystalline backbone of the High Himalaya extending eastward through Kashmir and Himachal Pradesh. This region, particularly the eastern end of the Kashmir Basin, marks the beginning of the patterns that give Himalayan geology its truly regional structure. A large crystalline uplift occurs east of the basin and west of the Zanzkar shear zone, with primarily granitic characteristics. To the east, in the Beas River Basin of Himachal Pradesh, the crystalline belt thrusts in a southwestward direction between the towns of Kulu and Rampur. North of Rampur in the famous Spiti Valley are extensive extrusions of quartzite similar to those found in the Kashmir Valley. Much of Spiti, though, is overlain by sediments, shales, pebbles, and conglomerates. The black shales, in particular, are common markers of the boundary between the High Himalaya and the Tibetan Himalaya, where Spiti is located. These dark shales and pebbly terrain give Spiti its characteristic stark and gray outlook.

To the east of Himachal Pradesh is a section of the western Himalaya commonly referred to as Garhwal or Kumaon—the terms referring to both a geographic area and a subrange of mountains. The Garhwal Himalaya consists of the 320-kilometer stretch of highlands between the Sutlej River and the Nepal border (defined by the course of the Mahakali River). Structurally, from north to south, this region includes the fourfold division noted earlier: the Tibetan trans-Himalaya, including famous Mount Kailas; the High Himalaya; the Lower Himalaya; and the Outer Himalaya or Siwalik zone. Of particular interest in this sector is the occurrence of the Vaikitra thrust north of the Baspa Valley, which splits the crystalline belt of the Garhwal Himalaya into two distinct parts.

The main central thrust is especially well outlined in the geology of eastern Garhwal and is easily discerned in the landscape looking northward from the middle mountains toward the high peaks. This juncture is evident, for example, when viewing Api Peak along the northwest Nepal–Garhwal border zone, and the famous pilgrimage sites of Garhwal, including those at Shivaling, Gangotri, and Badrinath, are overlooked by granite and phyllite peaks, which compose the crystalline belt of the High Himalayan zone. The upper stretches of the Garhwal waterways, including the Alaknanda and Bhagirathi rivers, carve deep gorges through the hard rocks of this region.

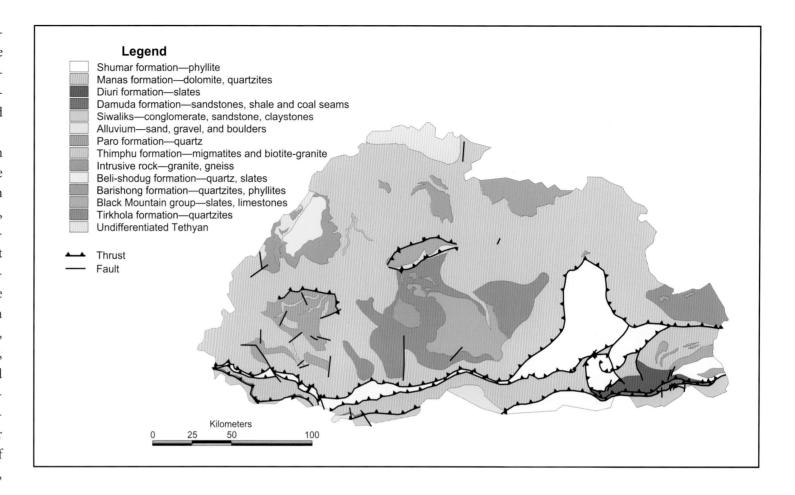

### GEOLOGY OF BHUTAN

Despite the pioneering work of Swiss geologist Augusto Gansser and the more recent joint surveys carried out by the Survey of India and Bhutan's Department of Geology and Mines, less than 30 percent of Bhutan has been mapped geologically. Mineral exploration began only in the 1960s. The tectonic structures of Bhutan follow the general pattern of the entire Himalaya, with less well-defined divisions between the High Himalaya and the Lower Himalaya and the presence of broad erosional valleys in the eastern parts of the kingdom. The northern border of Bhutan, in the trans-Himalayan zone, is marked by a series of marginal or axial mountains that protrude from the Tibetan Plateau.

Kagbeni village in lower Mustang occupies a transitional valley between the Tibetan Plateau and the High Himalaya.

along a tectonic fault separating the zone from the low-lying Outer Himalaya. As elsewhere in the Himalaya, the Siwalik foothills zone in Garhwal and Kumaon constitutes a region of considerable tectonic activity due to the faults and folding along the Himalayan frontal thrust.

### Central Sector

The central Himalayan region of Nepal provides some of the range's most extensive geologic records, due in great part to the ten-year survey begun in 1950 by Swiss geologist Toni Hagen. His field research took him to all parts of the kingdom, and his copious surveys and notes contribute to our baseline knowledge about the geology of this central Himalayan region. Geologists often accompany the summit assaults on Nepal's highest peaks and provide detailed surveys of them. All nine of Nepal's 8,000-meter peaks are located in the massive main central thrust zone. The Everest region, in particular, has seen a great deal of geologic exploration, beginning formally in the 1920s with expeditions originating on the Tibetan side of the mountain. It was not until 1952 that geologists first visited the southern side of Everest. The base of the mountain is composed of granites, which give way with altitude to gneisses, schists, and phyllites and then, near the summit, to limestone. The calcareous cap of Everest originates from the ancient Tethys seabed, which was once overlain by coral reef and sediments.

South of the High Himalaya is the more intricate geology of the Garhwal Lower Himalaya. The lack of fossil evidence leaves much of the stratigraphy a mystery, but recent tectonic investigations indicate sedimentary zones of limestone and sandstone separated by a crystalline belt. The Krol belt of limestone, slate, and sandstone, named after the Krol Mountains located near Simla, stretches between Simla and Nainital and is one of the most prominent structures in the Lower Himalayan zone. Near the town of Nainital are some of the oldest outcrops, and the relatively high mountains of the region lie

The great thrust sheets that compose the High Himalaya of the central region overshadow the Lower Himalayan zone, which occupies much of the terrain of Nepal. The Lower Himalaya, in turn, is divided into two main tectonic units: a lower sedimentary group and a higher, mostly crystalline formation called the Kathmandu unit. Each of these units is subdivided into numerous individual thrust sheets, so that overall, the geology of the Lower Himalaya in Nepal is quite complicated. Topographically, though, the Lower Himalaya is well defined by the Mahabharat Lekh Range, which runs parallel to the main boundary fault and forms the southern boundary of the zone, and by the central thrust peaks to the north. A few large basins intervene, notably the Kathmandu Valley and the Pokhara Valley, which are important settlement areas. Frequent rock types in the Lower Himalaya include granitized schists, gneisses, quartzites, and phyllites.

The low-lying Outer Himalaya of Nepal contains the Siwalik Range as well as adjoining areas of alluvial deposits. Much of the Nepalese Siwaliks is composed of conglomerates, with pebbles and boulders derived from eroded materials of pre-Tertiary age. The folds and thrusts of the Siwaliks generally dip to the north and indicate the north to south movement of the nappes atop the northward-drifting crustal plate. In the extreme southern boundary of the Himalaya, the Siwaliks disappear into the alluvial fans and sediment-filled plains of the Ganges. One of the chief features of the Siwalik region in Nepal is the presence of several large valleys composed of tectonic depressions overlain by sediments eroded from the general uplift of the mountains. These basins play an important role in the flow and flooding of rivers throughout the Himalayan foreland and in the distribution and accumulation of eroded sediments carried downstream.

*Eastern Sector*
The geology of the eastern Himalaya is least known because of the challenges posed by severe monsoon rains, the thick cloak of vegetation that covers the mountains up to considerable altitudes, and government restrictions imposed on travel and scientific research. One of the main differences between the geology of the eastern region and that of the western and central sectors is a series of cross-structures and sedimentary folds that make the distinction between the High Himalaya and the Lower Himalaya less clear. For example, the thrust boundary separating the two zones is not clearly disclosed in the strata of Bhutan, so the demarcation of the High Himalaya simply begins with the crystalline mass north of the sedimentary structures of central Bhutan. No clear thrust line is evident. The high elevations of the Great Himalayan zone are characterized by granite and gneiss peaks carved by glaciers into serrated ridges and cirque slopes.

The Lower Himalaya in Bhutan contains the prominent Black Mountain Range near Trongsa and a number of large river valleys whose wide floors distinguish them from the narrower gorges and valleys found elsewhere in the Himalaya. The broad valleys in Bhutan result from the erosive power of the rivers, with their high water levels from the monsoon rains, and the less resistant underbelly of sedimentary rocks. The bedrock of the Lower Himalaya becomes less metamorphosed to the east, so the rivers in eastern Bhutan tend to create wider valleys than in the western part of the kingdom. In contrast to the poorly defined thrust boundary between the High Himalaya and the Lower Himalaya, the main boundary fault separating the latter from the Outer Himalaya is well established in Bhutan. The low-lying foothills show severe tectonic deformation, with successive faults and thrust sheets along a west-east strike. As elsewhere in the Himalaya, this zone poses the most serious risk for earthquakes and other seismic action.

To the east of Bhutan, in the Indian state of Arunachal Pradesh, the geology of the Himalaya is largely unknown. Overall, the elevations are lower, with the exception of Namche Barwa (7,755 meters), which anchors the southward bend of the Tsangpo (Brahmaputra) River where it empties from Tibet into India. Across the Tsangpo River is the peak of Gyala Peri (7,150 meters), and the intervening gorge has only lately been explored. The steep gradient of the river as it cuts through this eastern sector of the Himalaya suggests very young mountains. In a remote section of the gorge, the river gradient is captured in a series of high waterfalls, whose presence has only recently been confirmed. Beyond the Brahmaputra region, major fault

》 Summer monsoon season, Langtang Valley.

and thrust zones separate the Himalaya from the adjoining highland regions in Myanmar and China.

The geologic formation of the youthful Himalaya has been a steady process, characterized not so much by sudden upheavals as by a relentless movement of the earth's crust and the steady rise of land where the continental plates of India and Asia collide. As the mountains have grown, so too have the forces of erosion acting on them to reduce their size. The rate of erosion or denudation of the Himalaya is every bit as impressive as its uplift. This geologic denudation, which takes the form of surface erosion and mass wasting (landslides, rock slips, or other mass movements of earth materials), is a natural and inevitable characteristic of the Himalaya. Indeed, the steep relief of the mountains, their youth and rapid rate of uplift (1,500 meters in the last 20,000 years), and the weathering potential of the monsoon climate all predict extremely high rates of denudation. Compounding these natural occurrences are the effects of human-related land clearing, which in severe cases accelerates soil erosion and contributes to slope stability problems.

## Climate

There is no single climate in the Himalaya; rather, the mountains create such diverse geographic circumstances that climate becomes kaleidoscopic, with each twist and turn in the terrain, every change in altitude and orientation to the sun, resulting in a plethora of individual climate segments. The major controls on climate in the Himalaya include latitude, altitude, and location relative to the Asian monsoon airflow. From north to south, the mountains cover a range of latitude greater than 8 degrees, spanning temperate to subtropical zones; in North America, this is equivalent to the span of the Appalachian Mountains from Pittsburgh to Atlanta. Moreover, the topographic barrier of the Himalaya permits the tropical climate zone to extend farther north in South Asia than it does anywhere else in the world. This factor is most pronounced in the eastern sector of the range, where the Brahmaputra Valley funnels warm air from the Bay of Bengal into the mountains toward Namche Barwa and northward into eastern Tibet.

Temperatures in the Himalaya vary inversely with altitude at a rate of about 2 degrees Celsius per 300 meters of elevation loss or gain. Due to the rugged terrain, wide ranges in temperature are found over short distances. Local temperatures also correspond to season, orientation of the land toward the sun, and size of the mountain mass. The seasonal differences are most pronounced in the northwestern regions of the Indian Himalaya and western Nepal, where the winter months are characterized by temperate or frigid weather. Since temperature is directly related to solar radiation, the mountain slopes that get the most direct sunlight also receive the most energy and heat buildup. This effect becomes more pronounced with increasing elevation. In the topography of the Himalaya, where steep-walled valleys are common, two facing slopes may be only a stone's throw distance from each other, but their opposing aspects produce significantly different weather. A southern exposure may well provide an additional month of growing season. The overall size of the mountain mass also influences temperature because it acts as a heat island and therefore influences the energy budget. The immense scale of the Himalayan peaks means that the summits create their own climate, which may be radically different from that of nearby plateaus or valleys.

One of the most influential factors affecting the Himalayan climate is the Asian monsoon. The monsoon is not a rain but a wind that carries rain in the summer months. The wind is triggered by enormous air pressure differences between Central and South Asia, which occur as a result of the differential heating and cooling of the inner continent and the surrounding oceans. In the winter, a high-pressure system hovers above Central Asia, forcing air to flow southward across the Himalaya. Because there is no significant source of moisture, the winter winds are dry. In the summer, however, a low-pressure system forms over Central Asia and pulls moisture-laden air northward. The wet summer winds cause precipitation in In-

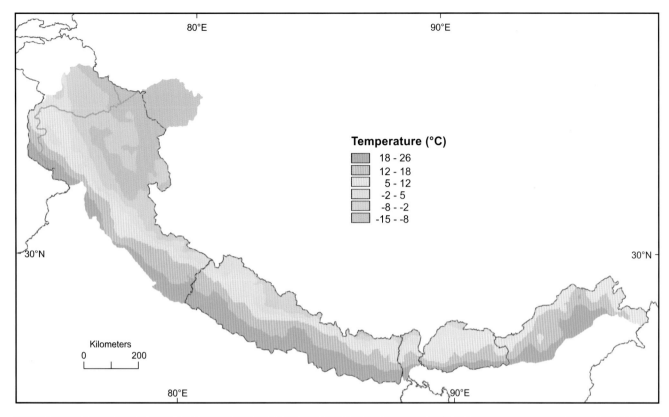

dia and along the tiered, southern slopes of the Himalaya. The water-laden monsoon air flowing north over the Himalaya is forced to ascend the mountains, where it cools, condensing and releasing its moisture as rain. This forced lifting of air over the mountains is called the orographic effect, and it creates a concentrated pattern of precipitation in the Himalaya.

The monsoon begins in the eastern sector of the range, in Arunachal Pradesh and Bhutan, at around the end of May. It then slowly moves westward, reaching Kashmir in the western Himalaya by late June or early July. As it moves westward, the monsoon also becomes drier. In the eastern region, the famous weather station at Cherrapunji in Assam records an annual rainfall of 10,871 millimeters, with a single-day record of 1,041 millimeters. This spot is the second-wettest place in the world, following Mount Waialeale in Hawaii, which receives an average annual rainfall of 12,344 millimeters. But whereas the rainfall on Mount Waialeale occurs throughout the year, Cherrapunji receives almost all its annual precipitation during the few monsoon months.

The monsoon precipitation drops progressively as one pro-

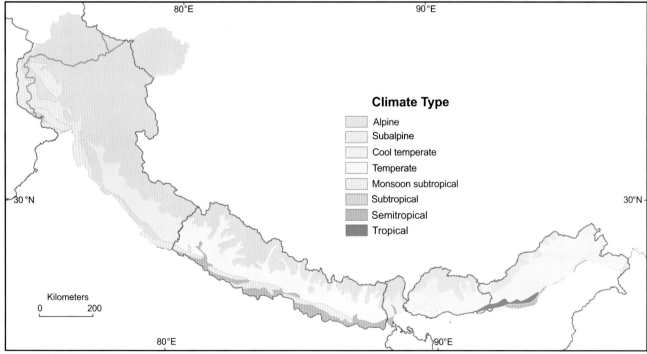

« HIMALAYA: MEAN TEMPERATURE AND CLIMATE

Altitude and latitude are the greatest influences on temperature in the mountains. The gradient of temperature change with altitude, known as the environmental lapse rate, coupled with latitudinal influences on temperature, produces generalized climate patterns in the Himalaya that range from subtropical to frigid alpine.

» LADAKH AND WESTERN HIMALAYA: PRECIPITATION

Precipitation (shown in centimeters) in the form of summer rain and winter snow has a pronounced seasonal and regional distribution, corresponding to the northward penetration of the summer monsoon and the rain shadow effect produced by the High Himalaya.

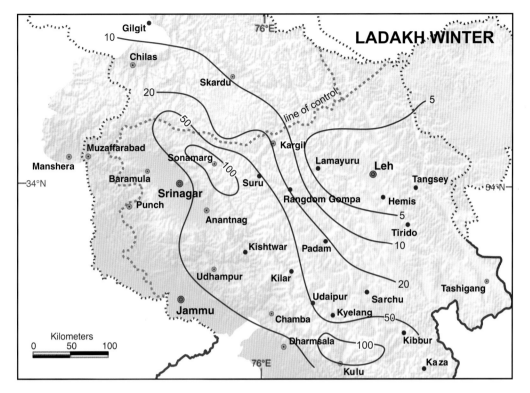
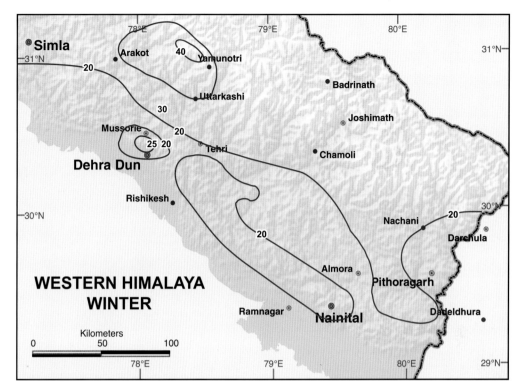
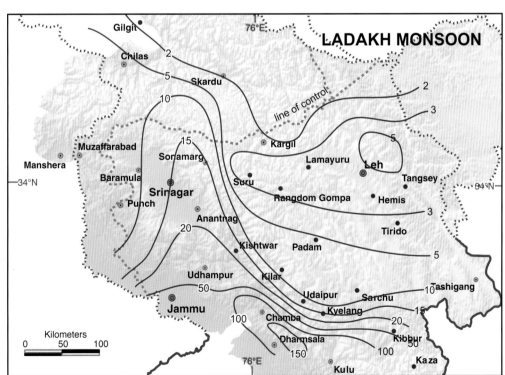
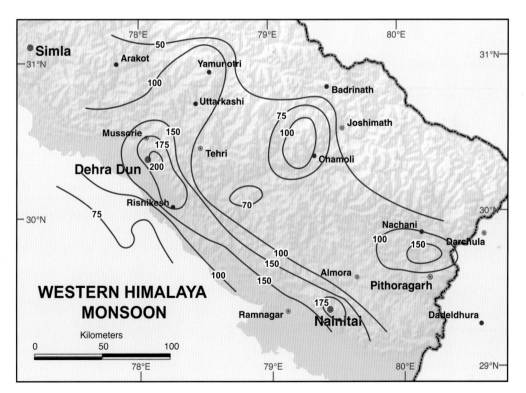

NEPAL: TEMPERATURE AND HUMIDITY

Altitude and landforms play key roles in the regional variation of both temperature and humidity in Nepal.

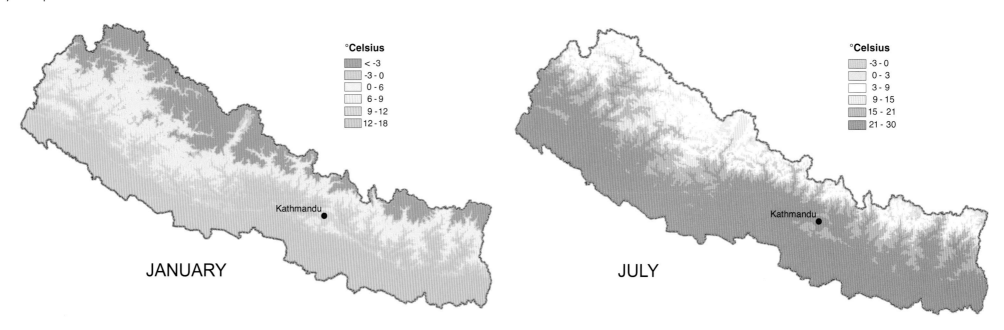

**NEPAL TEMPERATURE**

JANUARY

°Celsius
- < -3
- -3 - 0
- 0 - 6
- 6 - 9
- 9 - 12
- 12 - 18

JULY

°Celsius
- -3 - 0
- 0 - 3
- 3 - 9
- 9 - 15
- 15 - 21
- 21 - 30

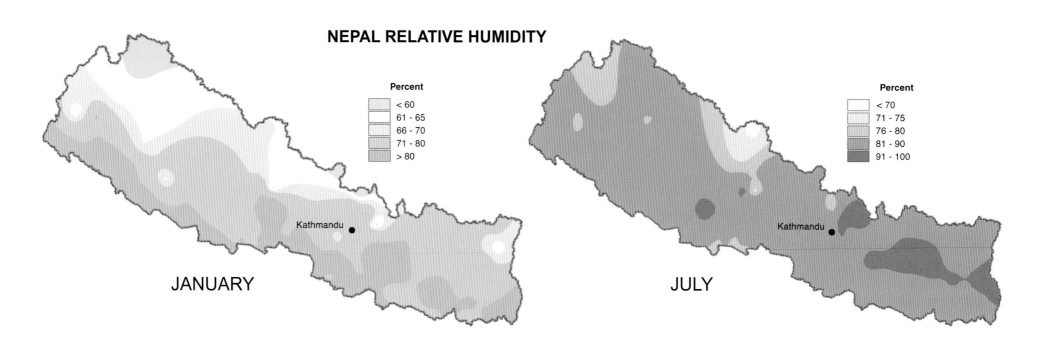

**NEPAL RELATIVE HUMIDITY**

JANUARY

Percent
- < 60
- 61 - 65
- 66 - 70
- 71 - 80
- > 80

JULY

Percent
- < 70
- 71 - 75
- 76 - 80
- 81 - 90
- 91 - 100

ceeds west, with annual receipts in Darjeeling of 3,122 millimeters; Kathmandu, 1,688 millimeters; and Jammu, 1,096 millimeters. There is a vertical gradient in rainfall amounts in addition to the longitudinal shift. An increase in rainfall occurs with altitude up to a maximum precipitation zone, which in the Himalaya occurs around 2,000 meters, after which it begins to drop again. The precise measurement of this gradient is difficult, in part because of the absence of recording stations at high elevations, but also because so many other factors, such as wind and solar direction, play critical roles in local temperature and precipitation accounts. In certain circumstances, however, the elevation factor actually supersedes the east to west longitudinal gradient.

When the wet wind from the south is carried over the High Himalaya, it has already lost much of its moisture, and the amount that remains is locked up as vapor when the air subtly warms as it descends onto the Tibetan Plateau. Consequently, the trans-Himalayan zone, in the lee of the high peaks, is dry. This is the so-called rain shadow effect. The barrier of the Himalaya results in startling contrasts. Precipitation in Nepal, for example, diminishes from 5,202 millimeters in Lumle, located on the southern side of Annapurna in central Nepal, to 174 millimeters on the north side of the same mountain. In Leh in Ladakh, which is located north of the main central thrust of the western Himalaya, annual precipitation is only 76 millimeters. In the eastern and central regions, it is possible to walk in only a few days from lush, wet forests to stark, cold, high deserts. Such transects make it clear that regional patterns of climate are often less important than local ones, which can vary in extreme ways over short distances.

Additional factors affecting the climate of the Himalaya include wind and glaciers. Mountains, which protrude into the high atmosphere, are some of the windiest places on the planet. They modify the normal circulation of air and create their own winds by setting up regional and local pressure systems. Mountain and valley breezes interlock in a diurnal circulation that can become so strong that it creates gale-force winds. In the Kali Gandaki Valley of central Nepal, the pressure gradients are especially powerful. The valley winds that blow through the Thakali village of Jomoson begin in the late morning and by noon may reach speeds of 80 to 100 kilometers per hour. These daily winds blow sand and grit through the valley in galelike conditions until late afternoon, when they taper off, leaving the air calm and the sky clear. Small valleys located below glaciers receive cool air moving downslope, which may rush like a torrent at times. Such glacial winds are generally thin, extending only a few hundred feet above the land, but the cold temperatures they bring to the places downwind can make them inhospitable indeed.

In the eastern and central Himalaya, most precipitation falls in the summer, and overall, the amount declines with altitude at elevations above 2,000 meters. The relatively low amount of rainfall received at the higher elevations partially

### NEPAL: TEMPERATURE AND PRECIPITATION (AVERAGE FOR 1990S)

| Weather Station | Precipitation (mm) | Temperature (°C) | |
|---|---|---|---|
| | | *Minimum* | *Maximum* |
| East | | | |
| Biratnagar | 1,943 | 19 | 30 |
| Dhankuta | 904 | 15 | 24 |
| Bhojpur | 1,041 | 13 | 21 |
| Center | | | |
| Janakpur | 1,361 | 20 | 31 |
| Kathmandu | 1,688 | 12 | 26 |
| Jiri | 2,235 | 8 | 21 |
| West | | | |
| Bhairahawa | 1,979 | 19 | 31 |
| Tansen | 1,634 | 16 | 25 |
| Pokhara | 3,957 | 16 | 27 |
| Far West | | | |
| Nepalganj | 1,480 | 19 | 31 |
| Surkhet | 1,431 | 16 | 28 |
| Dailekh | 3,098 | 11 | 26 |

Monsoon clouds rise with the warm morning air in a Helambu valley.

## AIR PRESSURE

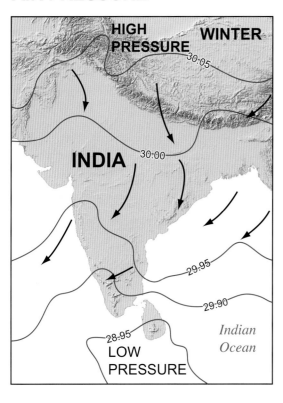
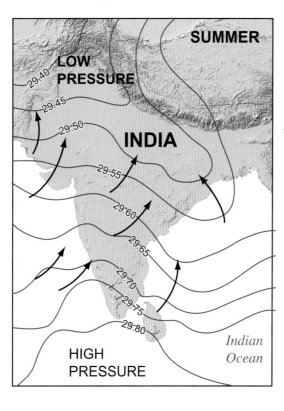

## RAINFALL

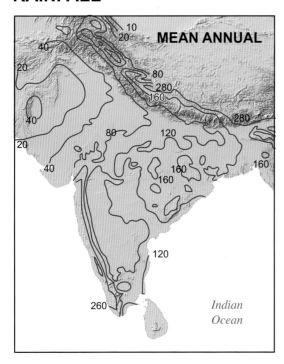
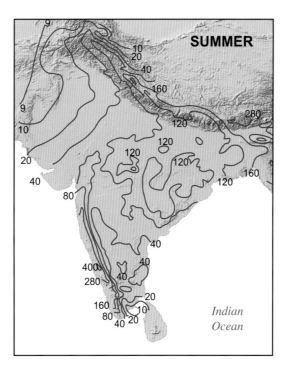

SEASONAL CLIMATE SYSTEM:
AIR PRESSURE (IN MILLIBARS)
AND RAINFALL (IN CENTIMETERS)

In the winter months, a high-pressure system builds over Central Asia, causing air to flow south. This is the winter monsoon, and it is a dry season due to the lack of moisture sources. In the summer, a low-pressure system builds over Central Asia, drawing air from the Indian Ocean. This moist air produces the precipitation associated with the summer monsoon.

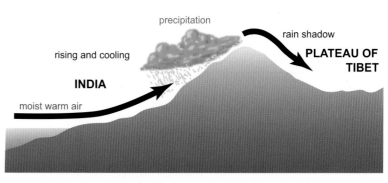

OROGRAPHIC PRECIPITATION

Moist air from the Indian Ocean flows north in the summer months across the Indian subcontinent and rises over the Himalaya. The rising air cools and condenses to form clouds and precipitation, which is most pronounced along the south-facing slopes of the range. As the airflow continues over the main crest of the Himalaya, it loses much of its moisture and warms as it descends. The result is a rain shadow, where precipitation rates are very low. This phenomenon results in the desert conditions found in much of the Tibetan Plateau.

57

explains why the glaciers there are not as vast as one might imagine. Their formation suggests that they are fed mainly by avalanches dropping snow from the peaks above, not directly by precipitation. In the western Himalaya and in the adjoining Karakoram Range, however, we find some of the world's largest glaciers. Their presence is attributed to the local topography, with its large basins, as well as to the fact that the western sectors of the High Himalaya receive considerably more precipitation from winter storms tracking from the west and producing locally severe snowfall. The significance of winter storms tapers off toward the east and is limited to only the highest summits in eastern Nepal, Bhutan, and Arunachal Pradesh.

Floodwaters, southern Nepal.

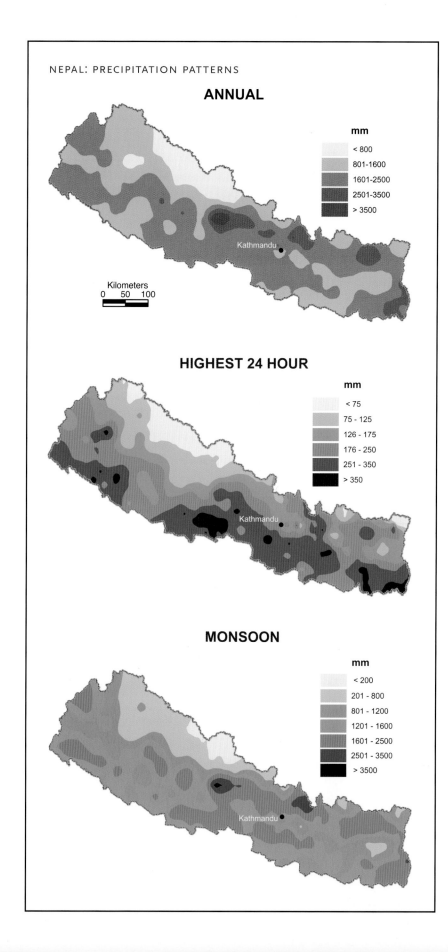

NEPAL: PRECIPITATION PATTERNS

ANNUAL

HIGHEST 24 HOUR

MONSOON

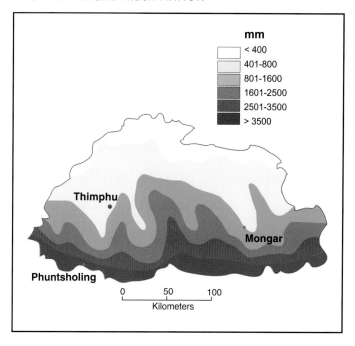

BHUTAN: ANNUAL PRECIPITATION

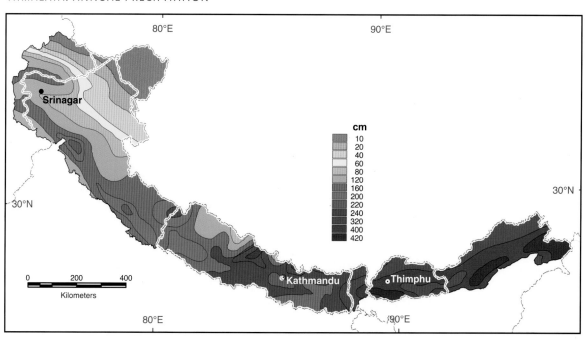

HIMALAYA: ANNUAL PRECIPITATION

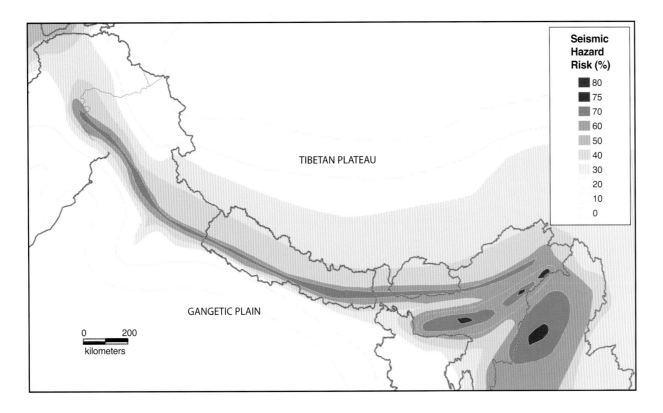

# Natural Hazards

The cycles of creation and destruction embedded in Hindu cosmology find a refrain in the landscapes of the Himalaya. Shiva, who is believed to have created the world, sits atop Mount Kailas with his consort, Parvati, who appears on earth as Kali, the goddess of destruction. From their vantage on the summit of holy Kailas, the deities look down on a vast mountain world that is caught between the opposing forces of uplift and erosion. India continues to drift northward, pushing against Asia; the thrust sheets that form the basement of the Himalaya grind inexorably upward; and the mountains, as a result, continue to grow. All the while, erosion acts to reduce the mountains. These two gargantuan forces—tectonic uplift and gradation—operate simultaneously to form the mountainous terrain, and both produce hazards that make the Himalaya a dangerous place to live. In some cases, these natural hazards are cataclysmic events—raging floods, glacial lake outbursts, earthquakes, landslides; other times, they are slow and insidious, such as the annual loss of soil by erosion, which renders agricultural fields less productive and threatens the stability of farm structures. By many accounts, the activities of humans—clearing forests, building roads, settling in disaster-prone areas—have contributed to the instability of the Himalayan environment. But the degradation caused by humankind, however significant for society, pales against the enormous power of natural events in the Himalaya.

Geologists refer to mountains as "high-energy" environments because they contain an enormous potential for such kinetic events as mass wasting, earthquakes, and floods. Mass wasting refers to the dislodgment of earth debris in a single, momentous action, such as a landslide. It is a natural phenomenon in the Himalaya—the dominant process in the formation of mountain slopes—and is common in the most geologically active zones and where precipitation is extreme. Sometimes, the results are spectacularly disastrous. In 1841 a huge chunk of earth fell from the west side of Nanga Parbat

## ⌃ HIMALAYA: SEISMIC HAZARD

The areas most likely to have seismic events occur along the axes of the major tectonic structures, mainly in the southern reaches of the Lower Himalaya and in the Siwalik foothills zone. The western extension of the hazard zone reflects the devastating earthquake that struck northern Pakistan in October 2005.

## » NEPAL: SEISMIC HAZARD

With almost daily occurrences of small seismic tremors and a large earthquake about every seventy-five years, the risk in Nepal is high. It is concentrated in the central and western zones, in a belt that includes the most densely populated regions of the country. Geologists predict that the next major earthquake (greater than 8.0 on the Richter scale) will occur by 2050.

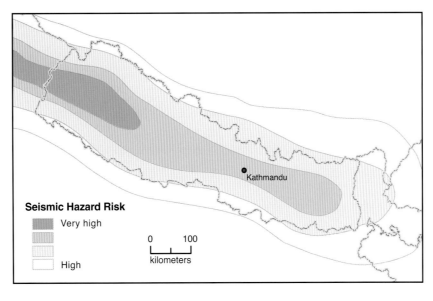

into the Indus River, damming it and forming a 50-kilometer-long lake. When the dam burst a few months later, people died in floods 150 kilometers downstream. Smaller landslides are common throughout the Himalaya, most notably after the monsoon, when the rains trigger erosion and landslips and the scars appear the freshest. In certain localities, where the land is intensively used, landslides are produced by the actions of people, but these are generally secondary to acts of nature. Many factors are involved in the mass-wasting process, including rock type and weathering, slope steepness, presence of fractures and sheer stress, and heavy rainfall, which can activate landslides. Where people disturb the slopes in major ways, such as by building roads and dams, the occurrence of landslides is often accelerated.

A special kind of mass wasting occurs at high elevations in the Himalaya when natural dams formed by glacial debris break loose and allow the impounded waters of the highland lakes to flood downstream. During the past half century, rapid melting of the glaciers has created a large number of glacial lakes, which are dammed by the terminal moraines of retreating glaciers. The dams are composed of loose till and unconsolidated rock and are thus easily breached when the force of the impounded water becomes too great or when a seismic disturbance disrupts the structure. The sudden release of this water is called a glacial lake outburst flood. The most famous of such events in the Himalaya occurred more than 600 years ago in Nepal, when a 10-square-kilometer glacial lake located behind Mount Machhapuchare burst and surged into the Pokhara Valley, raising the floor of the valley with over 5 cubic kilometers of glacial debris. More recently in 1985, a moraine-dammed lake at the terminal of the Langmoche Glacier burst above Namche Bazaar in the Mount Everest region. A huge chunk of ice dropped from a nearby summit into the lake, causing a tidal surge that broke through the moraine dam, releasing the lake water into the Bhote Kosi River. The flash flood raged downstream 40 kilometers, destroying farmland, thirty houses, fourteen bridges, and a hydroelectric power plant. This event sparked concern throughout the Himalaya about the hazardous poten-

The historic buildings of Kathmandu have survived earthquakes in the past but are at risk from future seismic events.

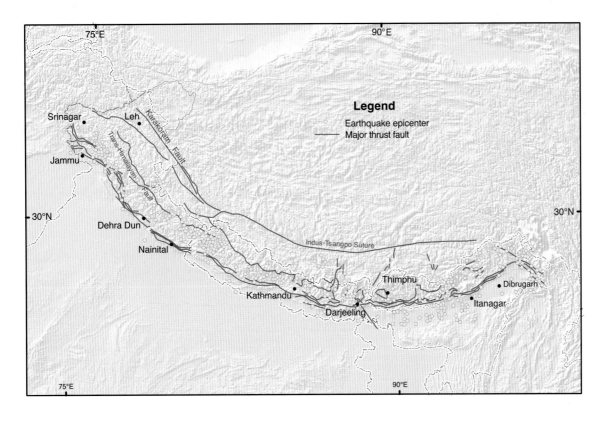

« EPICENTERS OF MAJOR EARTHQUAKES

Locations of the epicenters of significant earthquakes in the Himalaya during the past century are marked by yellow circles.

⌄ MAJOR EARTHQUAKES OF THE HIMALAYA

Red dots signify the epicenter locations of major earthquakes. Each dot is accompanied by the year of occurrence (below the dot) and its strength measured by the Richter scale (above). Note the correspondence between earthquakes and major fault lines and thrust sheet boundaries. These axes are the most earthquake-prone regions in the Himalaya.

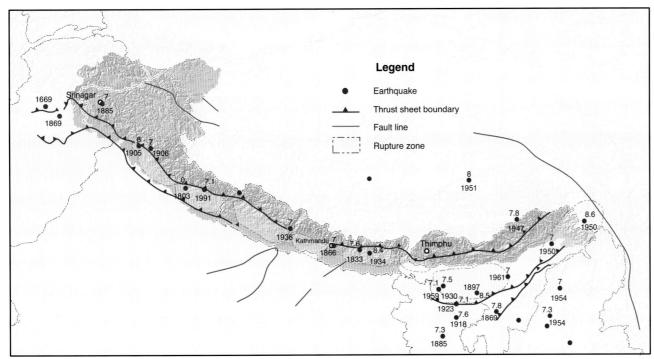

tial of glacial lake outbursts, and recent studies have identified numerous dangerous lakes throughout the region, including twenty-seven in Nepal alone.

The seismic hazard of the Himalaya produces the ever-present danger of earthquakes. They are a fact of life in the region and an almost daily occurrence. Eight major earthquakes measuring over 7.5 on the Richter scale have occurred in the Himalaya in the past century, and it does not take an earthquake that large to be disastrous. In 1991 an earthquake measuring 6.1 on the Richter scale struck the Garhwal region, killing more than 1,000 persons, destroying 30,000 homes, and ruining bridges and roads in a 60-kilometer radius. In Nepal alone more than 1,000 earthquakes ranging from 2 to 5 on the Richter scale are recorded every year. It is estimated that a truly big earthquake strikes the kingdom every seventy-five years; the last one occurred in the Kathmandu Valley in 1934 and killed more than 10,000 people. The region has been troublingly quiet during the last few decades, which means that a great deal of tectonic pressure is building, and current earthquake forecasters predict that a major event of catastrophic proportions will likely strike Nepal within the next fifty years. If it were to center again on Kathmandu, the consequences would

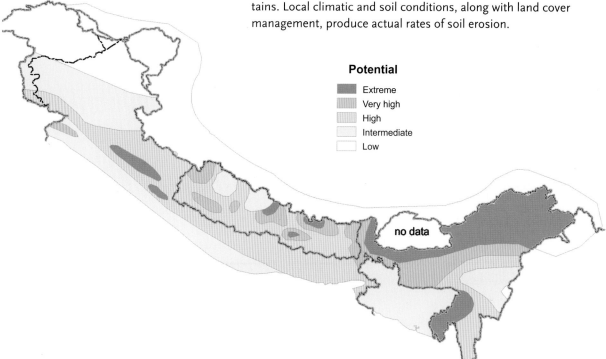

## SOIL EROSIVITY POTENTIAL

The kinetic energy of rainfall acting on soils of different types results in a soil erosivity potential. This generally increases in the Himalaya from west to east, following precipitation gradients, and is greatest along the southern flanks of the mountains. Local climatic and soil conditions, along with land cover management, produce actual rates of soil erosion.

**Potential**
- Extreme
- Very high
- High
- Intermediate
- Low

« Severe soil erosion and gullies endanger farmland in western Nepal.

be disastrous for the city. With an unstable sedimentary floor and poorly constructed buildings, Kathmandu ranks among the world's most vulnerable places for an earthquake disaster. Earthquake risk studies predict that such a tremor would destroy more than 60 percent of the city's buildings and result in more than 135,000 human casualties.

Seismic disturbances in the Himalaya may trigger sudden floods from burst lakes, whose effects, though disastrous, are generally contained by the river valleys below the lakes. More widespread flooding occurs as a result of the heavy monsoon rains and the high sediment loads carried by the rivers during periods of peak discharge. River sediments, originating mainly from natural mass wasting upstream, clog the lowland rivers and floodplains, silt up reservoirs and irrigation canals, and cause changes in river channels. Accelerated human population growth, forest clearing, and livestock grazing also contribute to higher erosion rates, which result in the loss of topsoil on cultivated lands and a decline in land productivity, as well as an increased sediment load in rivers downstream. Overall, however, the erosion damage caused by people is localized, and its contribution to regional flooding is minor compared with the natural rates of erosion. Nonetheless, lowland societies are

## » KATHMANDU VALLEY: LIQUEFACTION HAZARD

The Nepalese capital of Kathmandu sits in a valley where ancient lake sediments make up the valley floor. These materials move a great deal when there are seismic disturbances, and the ground actually liquefies in the case of severe shaking. Damage to structures and loss of human life will likely be significant in the case of a major earthquake, since most buildings in Kathmandu are not engineered to withstand this kind of stress.

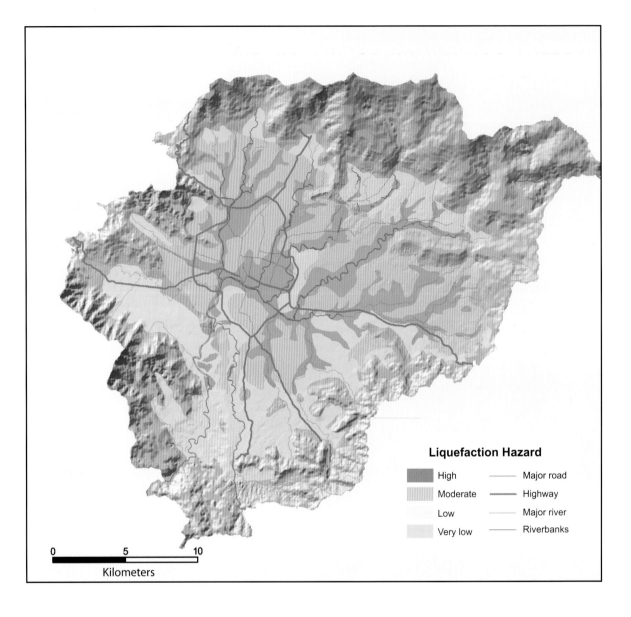

Landslide triggered by heavy monsoon rains carried away farmland, village homes, and a section of a mountain trail in Sikha, central Himalaya.

concerned and allege that the destructive activities of people living in the highlands produce the extreme floods that plague the Himalayan foothills and adjoining plains. The catastrophic floods in the Gangetic Plain actually result from a mixture of factors: the region's high and intense monsoon rainfall and the fact that more people are living in flood-prone areas, as well as heavy sediment loads in mountain rivers.

THE SPECTACULAR CHARACTER of the physical environment of the Himalaya—its remarkable topography and climatic extremes—is a fact of nature. Geologic uplift has produced the world's highest mountains and some of its most rugged terrain. The monsoon climate, in association with the mountains, creates some of the wettest places on earth. Geology and climate together produce a landscape of massive scale and ex-

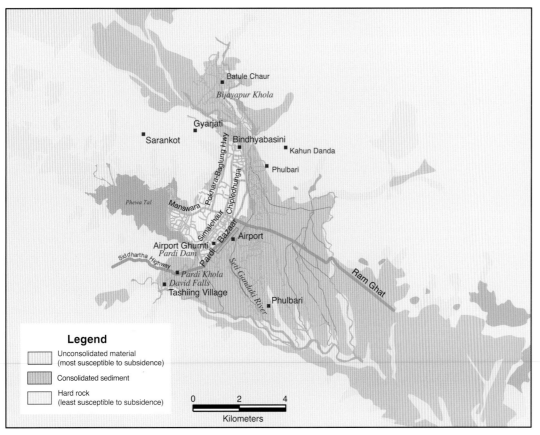

« POKHARA VALLEY: SUBSIDENCE HAZARD

Nepal's fourth largest city, Pokhara (estimated population 187,491, 2001 census), sits alongside lake Phewa Tal and expands into the adjoining areas of the Pokhara Valley. Much of the valley comprises unconsolidated materials and consolidated sediments, both of which have a high risk for subsidence, rockfalls, gully erosion, and flooding. This results in the regular occurrence of road blockages, bridge washouts, and waterlogged farm fields. The engineering required to overcome these hazards is costly and not always effective.

˅ Himalayan roads face the constant threat of landslides. Here, a 100-meter section of road in the Lahaul district slipped away, causing vehicular traffic to detour more than 500 kilometers.

treme physical events. The hazardous nature of the Himalayan environment and its propensity toward cataclysmic events such as earthquakes, landslides, and floods make the mountains a dangerous place to live. Himalayan societies have long understood these dangers and have developed ingenious ways of coping with the instability in their alpine world. As a result, the cultures we find in the range are every bit as diverse as the mountain environment itself. The trick to living in the Himalaya has always been to make good use of the natural resources it provides. Villagers intensively use these resources in their daily tasks of subsistence living. And nowadays, the mountains are perceived by the Himalayan nations and their societies as a frontier for new, large-scale economic development. The challenge at hand is to ensure that the natural resources of the Himalaya meet both needs in a sustainable fashion.

The Natural Environment   65

# PART THREE  Society

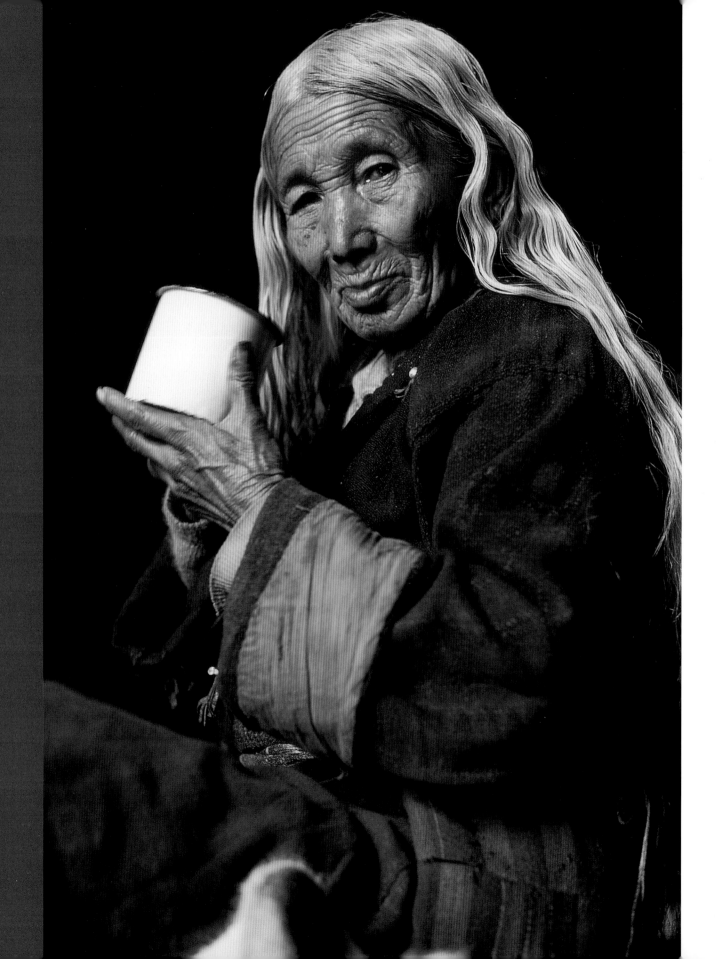

The Himalaya is home to the religious and cultural traditions of Hindu, Buddhist, and Islamic civilizations, as well as numerous tribes and ethnic groups. Within these diverse societies is a meeting of the spiritual and material, which defines a traditional order for mountain life. This order has led to the devout character of the Himalayan people, to the design of local communities, and to the establishment of historical theocracies in the mountains. It permits the inhabitants to display their distinctive cultural patterns and an astonishing array of lifestyles. And it shapes a landscape in which the textures of human society are interwoven in a traditional world of sacred places and powerful natural forces. But the demands of modern times and global trends have become dominant in many localities, instilling new forms of social organization and forging new appraisals of life that may create conflict as well as provide opportunity among the mountain communities. In this ever-changing world, Himalayan societies struggle between tradition and modernity.

The ancient religious books of India date the settlement of the Himalaya to the early Vedic times, before the age of Christ, although little is known about the actual origins of the initial inhabitants. The early chroniclers infused the mountains with a mythological outlook. The deity Shiva is believed to reside atop Mount Kailas, which is sacred to both Hindus and Buddhists, and many of the Himalayan peaks are said to be the abodes of other deities. The Himalayan rivers, especially the

» (part opener) Swayambunath Temple, Kathmandu.

« Bhotiya woman, Langtang Valley.

Painting at the entrance of a Buddhist monastery, Shermathang village.

Ganges, are revered as powerful spiritual places, and devout practitioners commonly make pilgrimages to their headwaters. The great rivers delimit a sacred geography that shows the Himalaya to be a truly celestial realm, within which native people find a spiritual order and guidance for life.

The religious bond that mountain cultures traditionally have with the mountains is manifested in the magnificent structures of the Hindus, Buddhists, and Muslims; in the ceremonial markers that adorn the landscape; and in the religious propitiation of the inhabitants. The Hindus understand the Himalaya to be the northern boundary of sacred India, or *Bharatavarsha,* just as the modern geologists consider it to be the tectonic delimitation of the subcontinent. The Buddhists, too, view the mountains as sanctified land, where places of legendary power contain a kind of knowledge that is fit only for those with the proper spiritual training. In the Indus Mountains in the west, the Muslims lay claim to long-held cultural territory that holds their religious history, mosques, and ancestral places. And the animistic traditions of tribal people all across the range find spiritual resonance among the summits and valleys, caves, forests, and rocky outcrops, which harbor deities of both good and bad intentions. These diverse religious beliefs embed the mountains within the cultural histories of the residents, thus creating a native geography that ultimately affirms the territorial rights and obligations of Himalayan societies.

« HIMALAYA: CULTURAL REGIONS

Three major civilizations converge in the Himalaya. Tibetan culture, based on monastic Lamaism and Buddhism, is dominant in the high-elevation regions and in the northern tier of the range. Indic culture, in which Hinduism is practiced, diffused northward from India and predominates in the southern tier of the range and among the lower elevations. Islam is found in the westernmost sector of the Himalaya. The high degree of cultural diversity produced by the convergence of these major civilizations is augmented by the numerous tribal traditions found throughout the mountains.

Society 69

## Early Civilization

The western Himalaya witnessed the migration of Aryan people as early as 2000 BC, most likely from the steppes of Central Asia. The Aryans settled the Indus Mountains in Kashmir before moving on to the plains of northern India. Little is known about the people they conquered, but the linguistic evidence suggests that the earliest tribes most likely originated in the far-off Dravidian societies of southern India. They lived autonomously in the mountains by hunting and gathering, possibly supplemented by slash-and-burn agriculture, before being conquered by the Aryans. An Indo-Aryan civilization emerged in the period 2000 to 1200 BC, as chronicled in the Aryan text the Rig-Veda. The later societies of the western and central Himalaya are documented in the Sanskrit literature of the Puranas and Mahabharata. These accounts describe a process of acculturation between the people of the mountains and those of the plains, most notably during the time of the Khasa people, when Hindu traits were assimilated into highland tribal life. The consolidation of Hindu influence in the Himalaya occurred in the fifteenth and sixteenth centuries, when coalitions of Rajput principalities were established across the range. A dominant Pahari culture emerged in the central Himalaya.

The high-altitude zone, meanwhile, was settled mainly from the north by Tibetan peoples. The seventh century witnessed the geographic expansion of Tibetan society into the western Himalaya, where it became dominant in Ladakh and Zanzkar. The Tibetan empire reached its zenith during this period, extending as far east as Turkestan, and formed the influential Ladakhi dynasty as a vassal state. The Tibetan forces also invaded the central Himalaya and, by AD 640, took control of much of Nepal as far south as the Kathmandu Valley. A series of Tibetan feudal states were established in the trans-Himalayan valleys, notably in such places as Dolpo, Mustang, and Khumbu, which continued as autonomous societies until

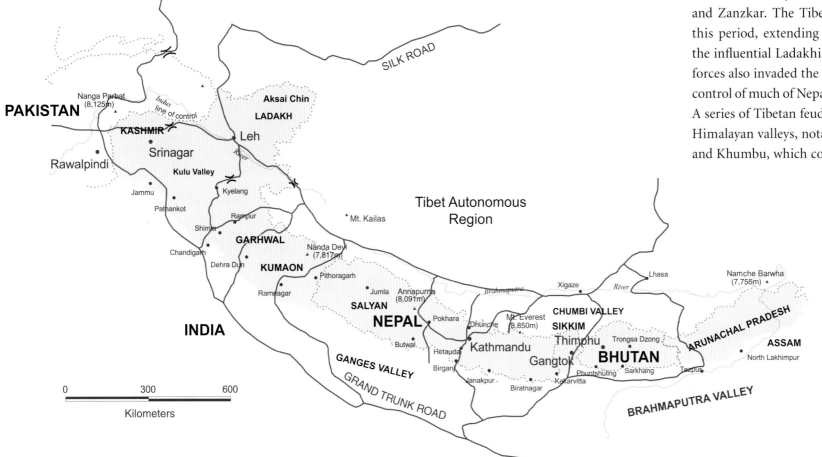

MAJOR HIMALAYAN TRADE ROUTES

The historic Silk Road in China was connected with the Grand Trunk Road in South Asia via a series of routes running south through the western Himalaya, mainly along the Indus Valley. Subsidiary Himalayan trade routes connected Tibet and India for purposes of the salt and grain trade, as well as for the exchange of other products. Many of these caravan routes followed the river valleys through the mountains, and several of them remain important trading routes today.

70  Illustrated Atlas of the Himalaya

fairly recent times. The Tibetan and Hindu influences intermingled most conspicuously in Nepal, whose society today is marked by a unique syncretism of Tantric Buddhist and Hindu practices. The scattered tribes living in the eastern Himalaya have retained much of their cultural autonomy and tribal traditions, residing largely outside the sphere of the major cultural invasions.

The medieval period in the Himalaya was marked by the consolidation of political power among the Hindu princes in the lower regions and of the Tibetan kingdoms in the northern mountains. Both subjugated the native populations, forcing villagers to produce crops, build infrastructure, maintain armies, and establish monuments and temples for the good of the petty kingdoms. Land taxes were introduced, and a rural aristocracy prospered amid widespread poverty. A monastic order was imposed in the areas under Tibetan control, which gave great power to the high-ranking religious clergy. In the Hindu areas, the local rajas governed as absolute sovereigns. In Nepal, the Hindu princes eventually succumbed to the expanding power of the king of Gorkha. The modern state of Nepal traces its origins to this Gorkha empire, which by 1815 had extended its territorial control across the Himalaya from Sikkim to Garhwal.

The convergence of Hindu, Buddhist, and, in the west, Islamic civilizations provided the cultural arena within which most Himalayan societies developed. The influences of these great traditions on mountain life are immeasurable, embedded in religious beliefs, family and community organization, artistic and folk traditions, and economy and politics. Between the middle of the nineteenth century and 1947, when India gained its independence, Himalayan society was also influenced by colonial powers. In the "Great Game" (as it is called in Asia), much of the mountain territory was coveted and disputed by the British, Russian, and Chinese empires. Each sought the mountains for its own imperial gain. The influence of the British on mountain life was substantial, for they directly controlled

Forest shrine to Kali, Garhwal.

Society 71

Hard-to-reach meditation caves in the predominantly Tibetan Buddhist trans-Himalayan zone. For the Tibetans, the Himalaya is filled with places of sacred power and spiritual insight.

large areas of the western and eastern Himalaya and had much authority over the internal affairs of Nepal and Bhutan. The British sent surveyors into the mountains to chart and map the territory and to identify their economic potential in terms of forests, agriculture, and minerals. Numerous skirmishes occurred early on between the British forces and the armies of the Himalayan kings, notably in Bhutan, Nepal, and Sikkim, heightening tensions among the competing rulers.

During their occupation of the mountains, the British opened up much of the range to colonial trade, drawing on the traditional trade routes—especially the lucrative Silk Route—while also establishing new ones. Timber resources in the mountain forests were exploited, trees were cut to provide materials to build railroads, and new regulations were imposed on villagers' use of colonial forest reserves. The British introduced new forms of agriculture, notably the apple orchards in Himachal Pradesh and tea plantations in Darjeeling. Where the British did not exert control directly, they did so through their influence on the local rulers. Gradually, as the colonial resolve of the British hardened, society and economy in the mountains became oriented toward the southern plains and even abroad. This new outlook continued after British withdrawal from South Asia in 1947 and the establishment of the modern nation-states in the Himalaya. Today, the independent countries of Bhutan and Nepal, as well as the adjoining Indian Himalayan territories, struggle with the enormous challenge of reconciling their pasts, including the colonial component, with their quests for a more prosperous future.

The famous chorten at Tabo, in Spiti Valley, is more than 1,000 years old.

# Population

One of the problems faced by modern Himalayan societies is the impact of a burgeoning population. In the past, human numbers were kept in check by high mortality rates and by cultural practices, such as polyandry, that acted to limit family size. Beginning in the latter part of the nineteenth century, though, the number of people living in mountain localities started to grow. Overall, an estimated 17 million persons resided in the Himalaya in 1890, when the first census was undertaken in the mountains. By 1950, the population of the mountains had reached 25 million, and by the beginning of the twenty-first century, it surpassed 50 million. The steady increase in population between 1890 and 1950 occurred unevenly, with colonial hill stations, road corridors, and fertile agricultural zones recording the highest rates of growth. Much of the concern about the negative impact of population growth, however, centers on the period after 1950, when the largest increases occurred.

In the last half of the twentieth century, more than 25 million people were added to the Himalayan landscape. Nepal witnessed some of the highest growth rates, with the population in several of its districts increasing more than 4 percent per year. The Indus Mountains, the lower elevations of Garhwal, and Sikkim exhibit very high rates of population growth.

HIMALAYA: POPULATION DENSITY

The population density in the Himalaya has increased along with overall population growth. This translates into increasing pressure on available farmland and forests, as well as on water resources. In the mountainous districts, where much of the land is sloped and therefore not conducive to agriculture, population densities are even more acute.

Family, Siklis village, Nepal.

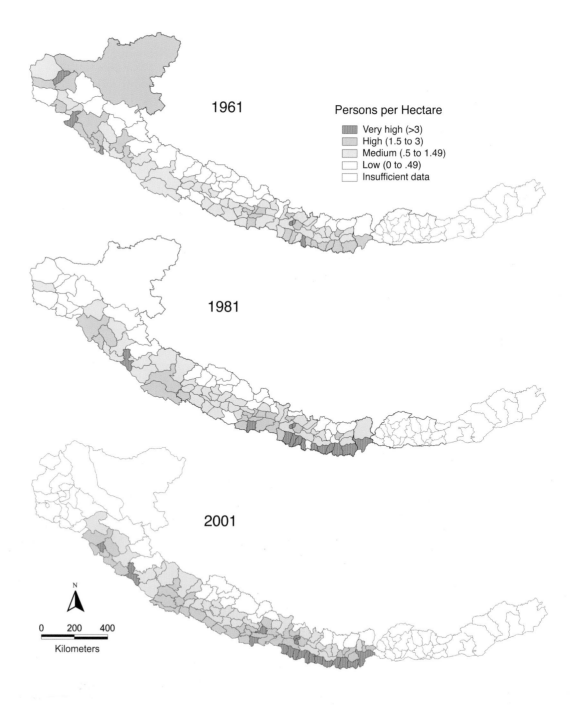

HIMALAYA: POPULATION GROWTH

Historically, population growth was slow in the Himalaya, with significant rates of growth reported in only a few scattered districts. In the contemporary period, growth was greater. Much of this was centered in the middle hills and in the tarai region of Nepal, where population growth rates exceeded 4 percent per year in some districts. The high rates of growth in the lowlands are partially a result of migration trends.

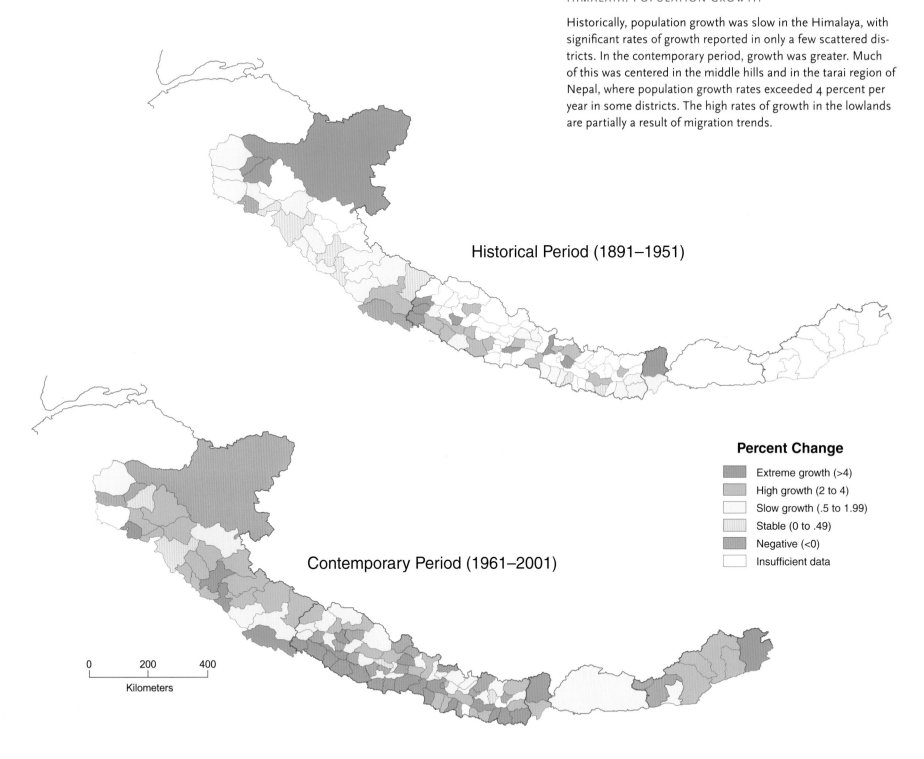

Historical Period (1891–1951)

Contemporary Period (1961–2001)

**Percent Change**
- Extreme growth (>4)
- High growth (2 to 4)
- Slow growth (.5 to 1.99)
- Stable (0 to .49)
- Negative (<0)
- Insufficient data

Bukarwal herder, Kashmir.

Ladakhi women.

### HIMALAYA: ETHNIC GROUPS

The great cultural diversity of the Himalaya is reflected in the numerous ethnic groups residing there. The main groups are located on the map, but there are numerous other small ethnic populations with their own languages, customs, and religious practices. In many cases, a cultural group is identified with a particular mountain or valley.

Newar shopkeeper, eastern Nepal.

Garhwali women threshing wheat, Sangla Valley.

Tibetan tribal women, Spiti.

Apatani woman, Arunachal Pradesh. (Photo by P. P. Karan)

Tharu farmer, western tarai.

Drukpa boy, Bhutan.

Gurung woman, Annapurna region.

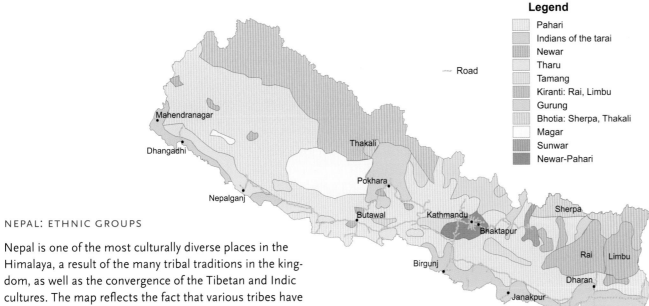

» NEPAL: ETHNIC GROUPS

Nepal is one of the most culturally diverse places in the Himalaya, a result of the many tribal traditions in the kingdom, as well as the convergence of the Tibetan and Indic cultures. The map reflects the fact that various tribes have long been settled in specific territories, and many groups are associated with particular mountains. One of the most famous ethnic groups, the Sherpa, live in the shadow of Mount Everest, which they call Sagarmatha. The Gurungs live near Annapurna, north of Pokhara, and the Magars are found mainly along the southern flanks of Dhaulagiri.

» NEPAL: LANGUAGES

The distribution of languages in Nepal closely follows the distribution of ethnic groups, with the various languages organized into four main language families. The dominant language families are Tibeto-Burman, which contains the Bhote languages spoken by people living in the northern part of the kingdom, and Indo-Aryan, which includes the national Nepali language.

Across the range, and especially in Nepal, the highest rates of growth occur in the outer foothills zone, where farmland is still available, and in the cities, which are expanding at alarming rates due to in-migration from the countryside. The population of the combined territories in the Indian Himalaya is over 20 million. Nepal has approximately 23 million inhabitants, and Bhutan has 2 million. The Indus region of northern Pakistan contributes an additional 4.3 million persons to the total. Important as these statistics may be, the Himalayan population is more than mere numbers; it is an extremely diverse collection of peoples of varied backgrounds. Maintaining this cultural diversity is one of the chief tasks for Himalayan societies.

## Culture and Ethnicity

With few exceptions, the ethnic diversity in the Himalaya is tied to specific geographic regions. The major religions of South Asia coalesce and maintain a unique expression in the western areas of Kashmir and Ladakh. The Indus Mountains and the northern portion of Kashmir are Muslim areas, Ladakh retains its Tibetan Buddhist heritage, and the southern part of Kashmir and the adjoining areas of Himachal Pradesh are predominantly Hindu. The largest ethnic group in the Garhwal and Kumaon regions is the Pahari, a term often ascribed to hill dwellers in general. In fact, the Pahari culture itself contains a great deal of internal heterogeneity, with members of different tribes and castes ascribing to distinctive lifestyles, manners of dress, and architecture.

Nepal exhibits perhaps the most remarkable mosaic of religion and ethnicity in the entire Himalaya. Hinduism is dominant throughout Nepal's lowlands and middle hills, and Buddhism prevails in the high elevations. In many places there is actually a Buddhist-Hindu syncretism, in which the religious practices, deities, and temples are shared by devotees of both traditions. Significant numbers of Muslims reside in the Nepalese tarai. These are mainly immigrants whose origins lie in India. Altogether, Nepal encompasses thirty-one different cultural groups and fifty-two languages. The dominant ethnic group in Nepal in the middle mountains is the hill caste population, which includes Bahun, Chetri, and Thakuri peoples, as well as the artisan castes. Other significant ethnic groups in Nepal include the Sherpa in the Mount Everest region, the Gurung living on the southern slopes of Annapurna, the Magars residing near Mount Dhaulagiri, the numerous Bhotiya (Tibeto-Burmese) clans residing mainly in the trans-Himalayan valleys, the Rai and Limbu tribes of the eastern middle hills, and the Tharu who inhabit the tarai lowlands.

Bhutan is overwhelmingly Buddhist, and its native Tibeto-Burman people, known as the Drukpa, share fundamental cultural traits. The Bhutanese language, the distinctive native dress, and the traditional arts and crafts are fostered by the central government, which promotes the idea of a distinctive Bhutanese nationality. Nepalese immigrants who settled in the Bhutanese lowlands in the early twentieth century are known as the Lotshampas and constitute the country's largest ethnic minority. Darjeeling and Sikkim also contain significant populations of Nepalese migrants, who have intermixed with the native Tibeto-Burman population. The largest ethnic minority in Sikkim are the Lepchas, who originated in Assam but have lived for centuries at the base of Mount Kanchendzonga. The eastern Himalaya contains a large number of tribes practicing animistic traditions. Some of them, such as the Monpa, Sherdukpen, and Khampti, have adopted Buddhism. The overall tribal character of Arunachal Pradesh reflects the great concentration of ethnic groups who maintain their own languages and traditions, such as shamanism, shifting cultivation, clan dress, and other forms of material culture.

The latitudinal diversity of Himalayan culture is enhanced by the vertical distribution of lifestyles. Rice farmers of the Himalayan valleys live in tight clusters of homes made from traditional stucco and thatch. Agro-pastoral people live higher up in the mountains in stone and slate-roofed villages, and they combine their grain farming with seminomadic livestock grazing. A common feature of these groups is their seasonal

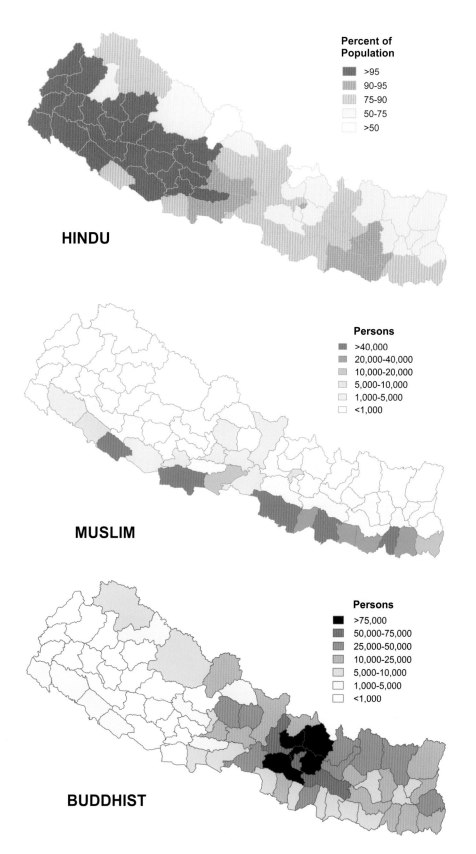

RELIGIONS IN NEPAL: HINDU, MUSLIM, AND BUDDHIST

The three maps show the distribution of the major religious groups in Nepal. The Hindu and Buddhist populations are long-standing, but the Muslim populations in the southern tarai districts are the result of the emigration of Islamic people from India since the 1970s.

Society 77

Mud and thatch homes in a Tharu village in the lowlands.

Flat-roofed adobe houses are common in the Tibetan-dominated villages of the arid trans-Himalayan zone.

migration to the high-elevation pastures, where they keep flocks of sheep and goats during the summer months. The trans-Himalayan valleys, in contrast, are settled mainly by Tibetan peoples. They live in flat-roofed, adobe structures and grow high-altitude grain in irrigated fields, raise potatoes, and keep herds of yaks and goats. Only one group of Himalayan people—the Bukarwal—remain fully nomadic. They move continually through the landscape of the Indus Mountains in Kashmir, ranging between highlands and lowlands as the seasons change. Some Himalayan valley groups, such as the Thakali and Newar, are best known for their trading and mercantile skills, while other highlanders are renowned as mountaineers (the Sherpa) or as mercenary warriors (the Gurkha soldiers recruited from the Gurung and Magar tribes).

The human diversity in the Himalaya is manifest in the architecture and temples; in the villagers' clothing, jewelry, and body tattoos; in the ceremonial practices, spiritual observances, and rites of passage; in the languages and dialects; and in the plethora of items that people place in the landscape. For example, the shapes and materials of homes reflect the practical needs of the occupants as well as their supernatural beliefs. In the cold and arid trans-Himalayan valleys, the houses are built of stone and have flat roofs, where the villagers can dry grain in summer and store firewood in winter. A huge array of house styles is found in the middle hills, ranging from the simple thatch and stucco dwellings of the Paharis and the oval-shaped Gurung homes to the elaborate, multistoried wooden structures of the Thakuris. Shamanistic traditions are common among the various mountain tribes, who live in humble bamboo and thatch homes.

Hand-woven textiles, which are prominent in the mountain villages, display particular colors and patterns that denote the residents' ethnic backgrounds. Some of the most exquisite tapestries in the world are created by those living in Ladakh and Spiti and by the tribes of Arunachal Pradesh. Women often wear their wealth as jewelry; the huge golden earrings and nose rings worn by Rai and Limbu women are indicative of the important cultural value of jewelry, which is a form of savings. Fluttering prayer flags, stone chortens and gateways, inscribed tablets in meandering rock walls, and copper prayer wheels turned by running streams all mark the distinctive Tibetan Buddhist realm. Elsewhere, sacred cremation places, sculpted stone deities, and pagoda-roofed temples denote the Hindu villages. Altogether, the mix of cultural traits in the Himalaya is stunningly complex and far-reaching, and it manifests the rich and varied human society in the mountains.

Novice monks, Bhutan.

Hindu temple pilgrims, Kulu Valley.

Chham temple dancers, Ladakh.

# Migration and Urbanization

People have always been on the move in the Himalaya. Long-distance trade routes crisscross the range, linking Tibet and India, and caravans have exchanged salt for grain across the rugged terrain for many centuries. Smaller mercantile routes also connect villages within more tightly bounded mountain settings, influencing the long-standing ties between highland and valley settlements. The seasonal migration of herders, which is a hallmark of high mountain culture, is an important adaptation to the environment, requiring people to move considerable distances on a regular basis. The herders trade animal products for grains and vegetables produced by the sedentary farmers. Marriage also traditionally requires one spouse to leave home and move to a new residence.

These age-old reasons for moving from place to place mean that migration is nothing new in the Himalaya, but the rate of migration and the reasons for it have changed considerably in the past several decades. Nowadays, many people move from

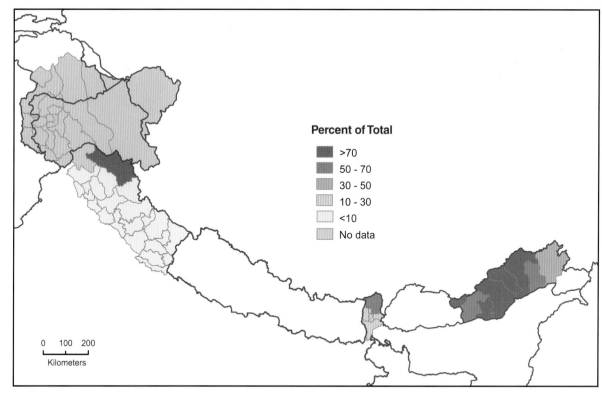

≪ INDIAN HIMALAYA: TRIBAL POPULATION

The Indian census groups indigenous peoples with a distinctive ethnic heritage in the category "tribal population" for statistical and policy purposes. The Indian Himalaya is one of that country's most diverse ethnic zones. In particular, the remote eastern mountain sector is a predominant tribal region, with a great variety of languages and religions.

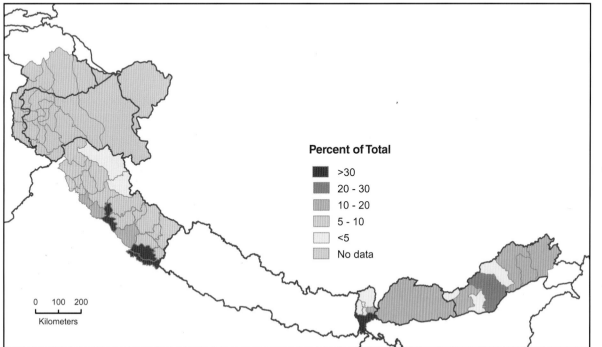

≪ INDIAN HIMALAYA: URBAN POPULATION

Although the majority of the population in the Indian Himalaya is still rural, the number of towns is growing, and so is the percentage of urban dwellers. This is especially notable where roads are common. It is also a trait of the larger valley districts located in the southern mountains. In the eastern sector of the range, where population size is relatively small, the percentage of urban population is greatly influenced by the towns that have sprung up at the base of the Himalaya, overlooking the Brahmaputra Valley.

Society  81

Migrant workers wait at a river crossing in Nepal.

Building construction in Kathmandu proceeds at a rapid rate as the city's migrant population expands.

their home villages out of desperation, because the land is no longer productive or because few economic alternatives exist in crowded rural settings. Food deficits in some rural areas force people to flee. Some move to other rural localities in the hope of settling new agricultural land; others make their way to cities with the idea of improving their lives. A growing number of individuals leave the Himalaya for places elsewhere in Asia, in the Gulf states, and farther abroad in the search of employment, education, or a Western lifestyle.

The Paharis of Garhwal and Kumaon in the western Himalaya have been moving onto the southern plains in significant numbers since the middle of the nineteenth century. Many of them were drawn to the employment opportunities offered by the British. The highland-to-lowland migration stream is more recent in Nepal, where migrants first began to flow out of the hills onto the tarai in steady numbers beginning in the 1960s, when the government initiated a land resettlement scheme in the lowlands. In 1971 the highlands of Nepal contained 62.4 percent of the country's population, but by 1991 that had declined to 53.3 percent, mainly a result of out-migration to the lowlands and cities. The tarai, which received 75 percent of that migration flow, also gained population because of emigration from India. The border between Nepal and India is open for both nationalities, and many undocumented Indians find it advantageous to live in the towns in Nepal's tarai zone. These migrants are turning the tarai into one of the most densely settled areas in the Himalaya and transforming its society into a hybrid of Nepalese and Indian influences.

The great majority of migrants settle in relatively new towns and cities. These places are growing at a very rapid rate, in most cases without sufficient infrastructure or planning. As recently as 1981, less than 10 percent of the Himalayan population lived in a town or city; by the turn of the twenty-first century, the urban population had doubled to 20 percent of

## TOWNS IN INDIAN HIMALAYA

Many of the larger towns located in the Indian Himalaya were established by the British during the colonial period as hill stations. However, the number and size of new towns are increasing along with population growth and modernization trends. The towns attract migrants who are seeking employment and an improved standard of living. All the major towns lie along important roadways.

### NUMBER OF TOWNS BY POPULATION

| State | Population (in Thousands) | | | | | | | | | | | | | Total |
| --- | --- | --- | --- | --- | --- | --- | --- | --- | --- | --- | --- | --- | --- | --- |
| | <5 | 5–10 | 10–15 | 15–20 | 20–25 | 25–30 | 30–35 | 35–40 | 40–45 | 45–50 | 50–55 | 55–60 | >60 | |
| Arunachal Pradesh | — | 5 | 3 | 2 | — | — | — | — | — | — | — | — | — | 10 |
| Sikkim | 7 | — | — | — | — | 1 | — | — | — | — | — | — | — | 8 |
| Darjeeling | 2 | 1 | 2 | — | — | 1 | — | 1 | — | — | — | — | 2 | 9 |
| Jammu & Kashmir | 24 | 20 | 6 | 1 | 2 | — | 3 | — | — | — | — | — | 2 | 58 |
| Himachal Pradesh | 34 | 9 | 5 | 2 | 4 | — | — | — | — | — | — | — | 1 | 55 |
| Uttaranchal | 26 | 13 | 7 | 5 | 4 | 2 | 2 | 1 | — | — | — | — | 5 | 65 |
| Total | 93 | 48 | 23 | 10 | 10 | 4 | 5 | 2 | — | — | — | — | 10 | 205 |

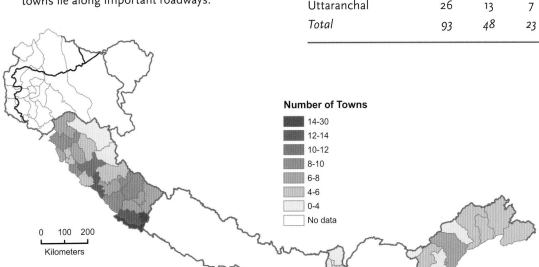

**Number of Towns**
- 14-30
- 12-14
- 10-12
- 8-10
- 6-8
- 4-6
- 0-4
- No data

### INDIAN HIMALAYA: RURAL POPULATION (1991)

| State | Rural Population | % of State Population |
| --- | --- | --- |
| Nepal | 22,212,624 | 91.40 |
| Bhutan | 17,567,000 (est.) | 80.00 |
| Jammu & Kashmir | 5,879,300 | 76.17 |
| Himachal Pradesh | 4,721,681 | 91.31 |
| Uttaranchal | 4,640,204 | 78.30 |
| Darjeeling | 903,859 | 69.53 |
| Arunachal Pradesh | 753,930 | 87.20 |
| Sikkim | 369,451 | 90.90 |

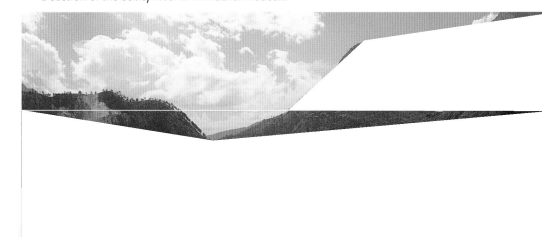

A new road, hydroelectric plants, and incipient industry have created a mini-urban corridor along a section of the Sutlej River in Himachal Pradesh.

## NEPAL: URBANIZATION (1952–2001)

People migrating from farming villages to towns in search of employment and education fuel the population growth in the urban areas of the country.

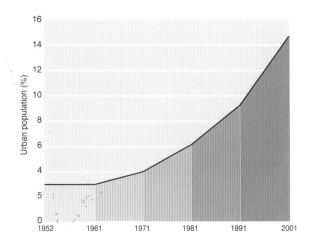

the total. In 1954 Nepal listed only ten localities with more than 5,000 persons; the 1991 census showed thirty-three such urban places, and in 2001 the number of designated towns and cities stood at fifty-eight. Nepal's urban population, meanwhile, increased from 9.2 percent in 1991 to 12 percent in 2001. Bhutan has a smaller urban population, less than 5 percent of the total, but that too is increasing. The urban population in neighboring Sikkim grew from 2 percent in 1950 to 16 percent in 1980. In the far eastern regions of Arunachal Pradesh, the population remains primarily rural, whereas the western part of the Indian Himalaya contains some of the range's largest cities (Srinagar, Dehra Dun, Shimla, Mussoorie, Nainital).

The lack of sanitation and waste management in Himalayan cities produces a public health concern that grows with urbanization. More than 70 percent of the people living in Nepal's Kathmandu Valley dump their garbage in the streets (where only 40 percent of it is collected) or along the riverbanks, where it piles up to create local disease hazards. The supply of drinking water is inadequate in most of the large cities and towns, and urban residents still rely on antiquated public wells and springs for potable water. The concentration of industry in some areas, which attracts migrants looking for employment, also contaminates the urban environment. Air pollution is now common among large towns, and rivers are often badly polluted. Kathmandu, which was relatively pristine through the 1970s, is one of the world's most contaminated cit-

Ziro valley and town, Arunachal Pradesh. (Photo by P. P. Karan)

## TOWNS IN EASTERN HIMALAYA

Urban settlements in the sparsely populated eastern Himalaya are restricted to river valleys and road heads.

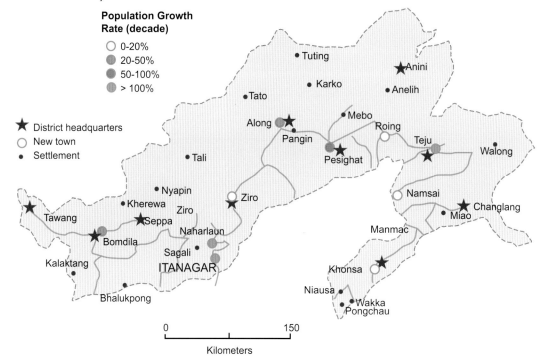

84  Illustrated Atlas of the Himalaya

ies. Its air is now thick with smog generated by factories and auto emissions, and its rivers are polluted and discolored by carpet factories and household waste.

Some Himalayan towns have ancient origins as pilgrimage centers or as capitals of feudal kingdoms. A few, such as Shimla and Darjeeling, developed under the British as hill stations. Most of the new towns, though, stem from regional economic development occurring in the mountains. Large development projects, such as hydropower schemes, require large administrative and labor forces that settle in the area and create new towns. Many people migrate to these towns to seek jobs in the new industries being created there. They work in factories, build roads, or undertake all sorts of menial labor. Most of the new towns are sprouting up along new roads. In Nepal, for example, such places as Dhankuta and Dharan in the east, Dhunche north of the Kathmandu Valley, and Daikekh in the far western region all owe their recent development to transportation inroads into the mountains. In the western Himalaya, which enjoys a relatively advanced road system, new towns are giving the mountains a decidedly urban and industrial look.

Iron bridge spans a river in the northwestern Himalaya.

## Transportation

Himalayan roads are engineering marvels, traversing mountain passes, fording rivers, cutting through the hard rock cliffs of canyons and steep ridges. The highest road in the world starts in Leh in Ladakh, rises 2,000 meters in 15 kilometers, crosses the Khardung Pass at 5,340 meters, and ends in the Indian army station located at the foot of the Siachen Glacier. The road was built to supply the army, but the public can travel on it as far as the Nubra Valley. Many of the newest high roads of Ladakh, Zanzkar, and Spiti in the western Indian Himalaya were built by the military for border defense, but most are now open for civilian traffic. They have made the remote valleys of the trans-Himalayan zone much more accessible for purposes of social and economic development. The roads that run north to south often must cross over difficult passes before they reach the plateau zone. For example, the towns of the Indus Valley are reached by road only after climbing the Zoji La from Kashmir or the Rohtang Pass from Kulu Valley. Both passes are closed much of the year due to heavy snow or avalanches. Across the Indus River in Pakistan is one of the world's great modern roads—the Karakoram Highway, which links the Grand Trunk Road of South Asia with the Silk Route of China. It traverses the 4,880-meter Khunjerab Pass and the magnificent Hunza Valley, covering some of the most rugged terrain in the world. Altogether, about 15,000 kilometers of roads have been built in the Indian Himalaya during the past several decades.

## NEPAL: ROAD NETWORK

The development of Nepal's hinterlands is tied to road development in frontier regions, including those that link China and India.

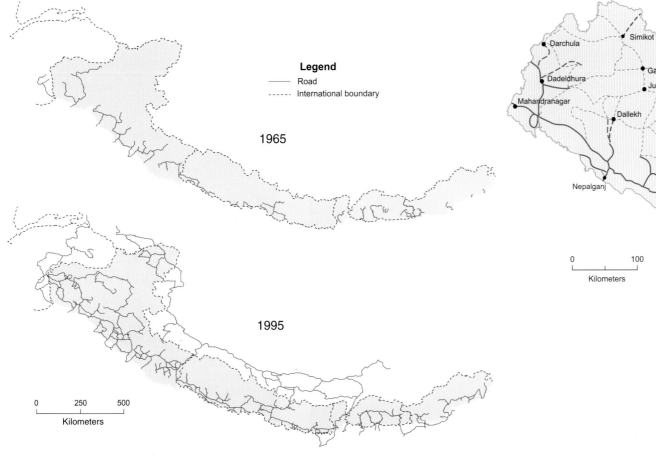

## HIMALAYAN ROADS

The period 1965 to 1995 witnessed a road-building frenzy in the Himalaya. In the western and eastern sectors, much of the road construction was targeted for military purposes, and many of the roads remained off-limits to civilians until the late 1990s. Elsewhere, the roads were part of national or regional economic development initiatives. The presence of roads means increased accessibility for purposes of market development and the delivery of social services. But roads also bring problems—notably, environmental instability. The road cuts in steep regions contribute to landslide problems and soil erosion. They also make it easier for illegal timber cutters to expand operations in the remote forests.

The first roads were built in Nepal beginning in 1953 and were limited to the tarai and the Kathmandu Valley. A road link between Kathmandu and the outside world was completed in 1956. By 1964, however, Nepal had only 289 kilometers of roads. Limited financial resources and the rugged mountain terrain hindered road construction throughout most of kingdom until the late 1960s and early 1970s, when a flurry of road-building efforts began. The Chinese and the Indians built many of the early roads in Nepal, both countries seeking to gain strategic advantage in the mountains. By 2000, the country had over 2,500 kilometers of paved roads, most of which are located in the middle mountains and in the lowland tarai. The roads serve Nepal's modernization interests, which require making its distant rural areas more accessible for social and economic development. Still, about a third of the country's population has no road access. Today, roads are being built into the heart of the High Himalaya, connecting Tibet with Nepal and India and opening remote valleys in the trans-Himalaya to the rest of the country. One of the most ambitious road projects will traverse the Kali Gandaki Gorge—the world's deepest valley and some of the harshest terrain in the Himalaya—connecting the remote Mustang district with the lowlands in the south and Tibet in the north.

The eastern Himalaya remains the least accessible part of the range. The maps of the area are notably empty. Bhutan began road construction in 1959 with the assistance of India, and in 1990 it had less than 1,500 kilometers of highway. The longest stretch of road (546 kilometers) is the East-West Highway, which connects the capital Thimphu with Tashigang. Much of the remainder of the country's roadway consists of a network of feeder roads that connects the district headquarters with the East-West Highway. Much of Arunachal Pradesh re-

86  Illustrated Atlas of the Himalaya

mains without roads. The major highways in the easternmost Himalaya are limited to the important valleys of the Subansiri and Brahmaputra rivers and lead to such towns as Seppa, Ziro, and Pasighat. As in the west, much of the road building in the eastern Himalaya had a military purpose, enabling the deployment of Indian troops near the border with China in a fractious tribal territory that is still claimed by China but occupied by India.

The problems encountered in Himalayan road building are formidable. In the first place, roads are costly to build. The engineering and material investment is great, and most Himalayan countries rely on donor assistance to build their roads. Most roads at some point cross a major river, requiring the construction of costly bridges. Many of the current bridges were built as temporary affairs and have exceeded their design lives. The heavy rainfall in the summer causes numerous landslides along the road alignments. Debris clearing, repairing undercuts of road embankments, replacing asphalt surfaces, and numerous other seasonal maintenance tasks are simply beyond the resources of the Himalayan countries. In Nepal between 1980 and 1993, environmental damage to roads resulted in US$34 million worth of repair work. As a result of the high maintenance costs, the roads in many mountain localities are in bad shape.

A number of alternative modes of transportation exist to augment the roads. Historically, the movement of people and goods across the Himalaya depended on trails and porters, and these still constitute the main way of getting around in most parts of the mountains. The plateau regions and wide valleys of the western Himalaya are crossed by pony caravans, which are a primary means of trade and travel in many places in Ladakh, Zanzkar, and Spiti, as well as in the valleys and plateaus north of the Himalayan crest in Nepal. The steep trails in the mountains, however, often cannot be traversed by pack ani-

Road workers, Bhutan.

» Bridges have improved considerably in the past few decades, making travel easier in parts of the Himalaya.

« Much of the Himalaya lacks roads, and the trails are too steep even for pack animals. Porters carry the heavy loads in these places.

Society 87

## NEPAL: AIRPORTS

Some of the remotest places in Nepal are served by air, with planes landing on dirt airstrips in rugged terrain. During emergencies, such airstrips may be the only lifeline to the rest of the country. Many of the mountain airstrips are seasonal and close down in winter or in the event of high winds or threatening storms.

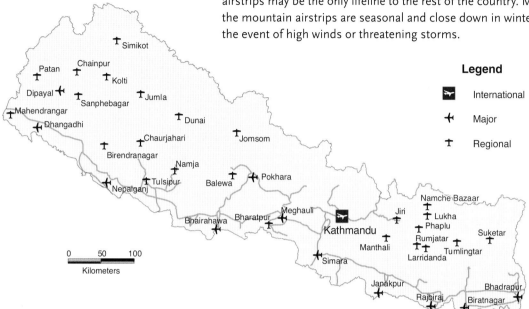

**Legend**
- ✈ International
- ✈ Major
- ✈ Regional

mals, so human porters have traditionally carried the goods on their backs. On some of the busier trails in Nepal, upwards of 1,000 porters may pass in a single day. Suspension bridges are an important feature of the foot trails and mule tracks, allowing the safe crossing of rivers and gorges. In recent years, the importance of ropeways has increased. The first ropeway in Nepal was built in 1927 to ferry goods into the Kathmandu Valley. Since then, numerous large and small ropeways have been constructed in the Himalaya, greatly reducing the burden of portering heavy goods up steep inclines in heavy-traffic areas.

The steep terrain of the Himalaya has not allowed much in the way of rail transport. Two narrow-gauge railways in Nepal link border towns with India and are used mainly for cargo traffic. Small, narrow-gauge railroads built by the British still go to Darjeeling and to Shimla in the Indian Himalaya. The so-called toy trains that move on these lines are used exclusively for passenger traffic, mainly people on holiday. A railroad goes as far as Pasighat in the eastern Himalaya and is used for mili-

Satellite dishes powered by small hydroelectric installations bring the outside world into remote mountain villages.

Helipad in the Himalaya.

88   Illustrated Atlas of the Himalaya

tary transport as well as for civilian purposes. Air transport plays a key role in many mountain localities, both for passenger movement and to ferry food and other necessities across impossible terrain. In Nepal, the construction of small, grassy short takeoff and landing airstrips provided a vital lifeline to many villages. Air transport began in 1950 with the construction of the airstrip at Kathmandu, and in 1958 the Royal Nepal Airline Corporation was established. Since then, the number of airstrips in Nepal has increased to over forty, and the fleet of private airlines has mushroomed, mainly to serve the tourism industry. Bhutan, in contrast, has only one airport at Paro, but numerous helipads exist across the country, which are used mainly for emergency or governmental purposes.

## Communications

The remote and scattered population of the Himalaya, combined with the lack of motor transport, makes the need for communication especially vital among the mountain communities for purposes of human development and participation in government. Large areas of the Himalaya still are not covered by postal services. Buses are used along roadways to dispatch mail, and air deliveries reach the remote airstrips. Otherwise, villagers rely on the unscheduled services of volunteer or government-paid runners to obtain printed mail. A few private delivery services, including DHL and UPS, now operate in some of the capital cities of the Himalayan region, providing express worldwide deliveries. These are costly, however, and serve mainly the business sector.

Telecommunications operating in the Himalaya include telephone, telegraph, wireless radio, and satellite transmissions. These are common in the cities, along with television programming, but the rural areas, particularly those with no electricity, remain without adequate communications. For example, only fifty-five of seventy-five districts in Nepal are served by telephone, and the entire country of Bhutan has fewer than 3,000 working telephone connections. The communications sector is more highly developed in the western Himalaya, especially in Himachal Pradesh, which is better served by electricity and by government as well as private telecommunications exchanges. Satellite telephones and television dishes are still rare in the mountains but can be found where electricity is available. Small hydropower stations provide electrical power, and parabolic dishes bring a global system of television programs to the remote villages. Most of the Himalayan towns and the villages along the roads are served by print media in national and vernacular languages, as well as English in the large cities. The Indian districts are served by the national Hindi newspapers, as well as by local tabloids. Nepal has numerous small weekly newspapers, many of them aligned with political parties; there are also the national private daily papers, *Kantipur* and *Kathmandu Times,* and the government

Communications tower, Stok Valley, Ladakh.

COMMUNICATIONS IN BHUTAN: TELEPHONE SWITCHING NETWORK

Traditional telephone networks in Bhutan, as well as elsewhere in the Himalaya, do not reach very far into the remote areas. These systems are gradually being augmented by wireless and satellite-based communications systems.

Farmer, central Himalaya.

Landless farmer, Nepal tarai.

papers *Gorkhapatra* and *Rising Nepal*. Bhutan is served by the weekly national newspaper *Kuensel*, which is published in Thimphu.

## Human Development

The United Nations' *Human Development Report* provides an index of variables (life expectancy, education, income) that, when taken together, measure the development of human society. In the UN calculations for 162 countries, the Himalayan region fares badly. Nepal came in at 129, followed by Bhutan at 130. India and Pakistan rated a bit better, at 115 and 127, respectively, but those countries' Himalayan territories contain some of their poorest districts. The UN calculations, which emphasize material wealth, are contested by Bhutan, which proposed its own "Gross National Happiness Index" in its *Human Development Report 2000* submitted to the United Nations. In the Bhutanese view, the goal of development is a happy society, which must consider spiritual and emotional factors, as well as material ones. Nonetheless, the common UN indicators of income, education, access to health services, and life expectancy represent worthwhile, if not exclusive, goals for national development, even in Bhutan.

### POVERTY AND EMPLOYMENT

Although the bulk of the Himalayan population remains agricultural, the traditional farming systems cannot absorb the growing workforce. The result is greater poverty. Therefore, the challenges of development in the region include the creation of alternative economic opportunities in the private non-farm sector. Currently, the region's income levels are among the lowest in Asia. Nepal's per capita gross national product (GNP) in 1998 was US$200, ranking it alongside the poorest countries in Africa. Over half its population lives on less than a dollar a day. India's overall per capita GNP is US$370, but its Himalayan region includes some of its poorest places. Sikkim, for example, reports a state per capita gross domestic product (GDP) of around US$215, while Arunachal Pradesh reports per capita incomes less than US$200. The per capita GNP reported by Bhutan is higher, at US$594; Bhutan also believes that when natural capital (forest, water, soil) is included, the per capita wealth in Bhutan increases to US$16,500.

Although the official overall unemployment rate in the Himalaya is quite low—less than 5 percent—its total workforce is underutilized by about 50 percent, due in part to the seasonal nature of farmwork. This latter figure is more important in modern times, because more people are seeking to earn a living in the towns and industries. In Nepal, the active workforce includes 22 percent laboring in services and less than 14 percent in manufacturing. The workforce participation of women in Nepal has almost doubled since 1971, while that of men has decreased. Nonetheless, poverty is alarmingly high throughout the country. In 1996 the percentage of the population living below the national poverty level ranged from 23 percent in the urban areas to 56 percent in the mountains. Agriculture, meanwhile, has largely stagnated, and many mountain districts are now experiencing a food deficit. Nepal's overall sluggish economy has led to slow income growth (1.4 percent per annum during the past twenty-five years) for everyone. The underemployment problem is compounded by labor migrants from India, who fill menial jobs as well as entrepreneurial roles. In Bhutan, the GDP grew at a rate of 7.3 percent per annum during the 1980s and 5.9 percent between 1990 and 1998. The mining, manufacturing, and energy sectors of the national economy contributed significantly to this growth, ranging from 4 percent of the national GDP in 1980 to 25 percent in 1998. These increases have led to new jobs in Bhutan's industrial sector.

The economic prospects in the western Himalaya vary considerably from one place to another. The decades of instability in Kashmir have led to economic stagnation in all sectors of the economy and to serious levels of income poverty throughout the state, as well as other forms of human misery. Himachal Pradesh and Uttaranchal, however, have fared better. The rural sector in Himachal Pradesh has invested heavily in commercial agriculture, notably orchards, which provides work in the fields as well as in the associated fruit-processing industries. The hydroelectric schemes in both Himachal Pradesh and Uttaranchal provide employment for menial workers as well as for technical and engineering staff. And tourism plays an important role in the rural service economy of both states. Throughout the Himalaya, cottage industries such as handicrafts, papermaking, textiles, and food processing are promoted as sustainable sources of livelihood for the rural mountain population.

With the growth of urban places, the prospects for employment in the industrial, service, and trade sectors have increased in some areas. Only small proportions of town dwellers are engaged in household industries (in the Indian Himalaya, this ranges from 0.3 percent in Arunachal Pradesh to 4 percent in Sikkim), despite its heavy promotion by government policies. The largest concentration of urban workers in the Himalaya is in trade and commerce and in other service sectors. These categories account for two-thirds of urban workers in the Indian Himalaya. Similar rates prevail in Nepal. Impediments to urban employment throughout the range include the lack of training and low educational levels, as well as the low wages paid to employees. With continued high rates of rural to urban migration, and with most immigrants moving to the cities in search of jobs, employment generation becomes even more im-

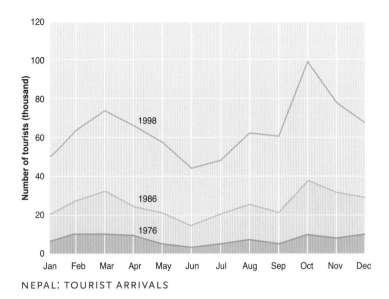

NEPAL: TOURIST ARRIVALS

Tourism in Nepal is seasonal, with spring and fall being the best times to visit the country. Civil unrest since 2000 has led to an alarming drop in the number of tourists, jeopardizing the economy.

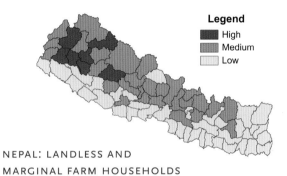

NEPAL: LANDLESS AND MARGINAL FARM HOUSEHOLDS

The problem of landless farmers occurs where population densities are high and agricultural area is limited. This is particularly acute in places where the rural elite controls a significant proportion of the available farmland. In Nepal, the western districts report high rates of landlessness; these are also some of the poorest districts in the country. Landlessness has become a concern in some of the lowland tarai districts as well, which have experienced high rates of immigration in recent decades.

Tourism provides employment in many parts of the Himalaya.

portant. In Sikkim, where people increasingly seek livelihoods in the cities, the poverty rate increased from 36 percent in 1988 to 41 percent in 1994, highlighting the need for job creation in towns.

## EDUCATION

Historically, educational levels in the Himalaya have been low, and they continue to be among the lowest in the world; however, recent decades have seen significant improvements in classroom enrollment and adult literacy. This change recognizes the fundamental role that education plays in human development. In Nepal, the adult literacy rate increased from 8 percent in 1961 to 45 percent in 1997. In a recent survey, parents in Nepal unanimously ranked education as the top priority for their children's future. Bhutan has witnessed similar strides in education, with its adult literacy rate increasing from 10 percent in 1970 to 40 percent in 1994. Still, less than half the Nepalese and Bhutanese populations can read and write, and education remains a top priority for both countries. Gender inequity also needs to be addressed, along with the efforts to improve overall literacy. In Bhutan in 1990, the boy-girl ratio in primary school was 61-39. This has improved somewhat, but women are still seriously underrepresented in schools at all levels. In Nepal, the gender imbalance is reflected by the fact that the literacy rate is 62 percent for men but only 28 percent for women. This discrepancy is even greater in the remote mountain villages. Some of the highest educational levels are reported in the Indian Himalaya, where schools and teachers are more numerous. For example, Sikkim's adult literacy rate is almost 70 percent, and about 83 percent of all children aged six to seventeen attend school. The Indian region's lowest literacy rates

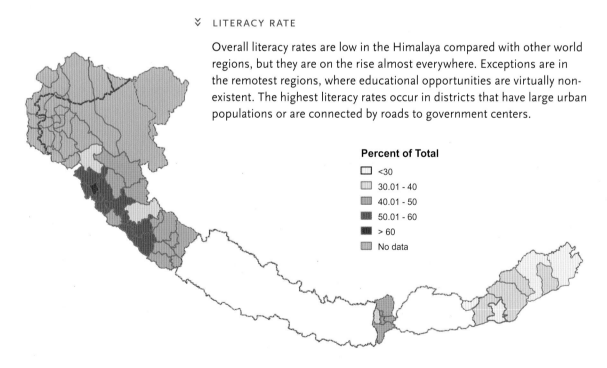

### ⌄ LITERACY RATE

Overall literacy rates are low in the Himalaya compared with other world regions, but they are on the rise almost everywhere. Exceptions are in the remotest regions, where educational opportunities are virtually nonexistent. The highest literacy rates occur in districts that have large urban populations or are connected by roads to government centers.

**Percent of Total**
- <30
- 30.01 - 40
- 40.01 - 50
- 50.01 - 60
- > 60
- No data

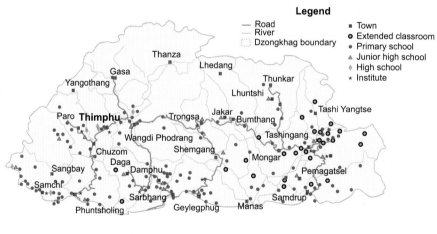

### ⌃ BHUTAN: EDUCATIONAL FACILITIES

The literacy rate in Bhutan remains low (41 percent), reflecting the lack of opportunities for formal education in many rural areas. The government schools are supplemented by monasteries that provide young people with instruction in reading and writing, as well as religious training.

are among the tribal groups and where urbanization levels are also low. Arunachal Pradesh, for example, reports literacy rates as low as 9 percent in the West Siang district.

The expanding demand for schools means that the institutional infrastructure of education must be developed. Currently in the Himalaya, there are too few teachers and schools. Bhutan has always faced a shortage of teachers, which it meets in part by employing foreigners, mainly Indians, in its schools. The teacher training institutes in Bhutan, however, are expanding their capacity to meet the new demand, and the number of graduating teachers increased from 487 in 1998 to 713 in 1999. The public schools in Nepal are also poorly equipped, and the teachers often do not receive proper training. As a result, the quality of public education is sadly lacking. This has led

School facility in the tarai.

» KATHMANDU VALLEY: WATER QUALITY

The rivers and streams flowing through the Kathmandu Valley have historically provided valley residents with water. These sources, however, are becoming increasingly polluted; the highest contamination rates occur closest to the city, but even the outlying valley has experienced diminished water quality. The water supply for Kathmandu is now augmented by large water diversions from sources outside the valley, among the high mountain streams to the north.

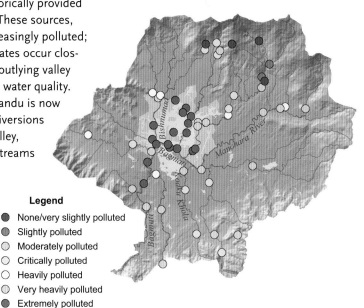

**Legend**
- None/very slightly polluted
- Slightly polluted
- Moderately polluted
- Critically polluted
- Heavily polluted
- Very heavily polluted
- Extremely polluted

⌄ BHUTAN: HEALTH FACILITIES

Government-sponsored health units are scattered across the countryside of Bhutan, but many of these are poorly equipped and lack adequately trained personnel. As a result, many people go without modern health care and rely on traditional village healers.

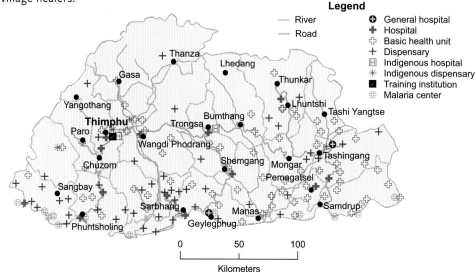

**Legend**
- River
- Road
- General hospital
- Hospital
- Basic health unit
- Dispensary
- Indigenous hospital
- Indigenous dispensary
- Training institution
- Malaria center

⌄ NEPAL: ACCESS TO POTABLE WATER

Drinking water supply systems in the villages rely on local springs and streams. The water is diverted to households by plastic pipes, which may carry water aboveground for several kilometers from its source. Reliable and safe drinking water is a high priority in many rural areas, where declining water tables or pollution may threaten the potable water supply. This is a particular problem in the most densely settled rural localities.

**Percent of Population**
- >75
- 60-75
- 45-60
- 30-45
- <30

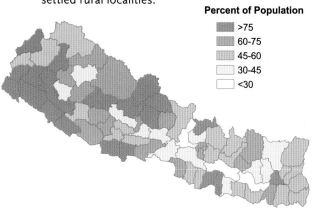

to an increasing number of private schools, including costly boarding schools, and many parents choose to make financial sacrifices elsewhere in order to send their children to better schools.

In sum, educational improvement across the Himalaya requires better access to educational facilities and properly trained and equipped teachers. Socially disadvantaged and minority groups must be targeted, and a set of educational goals must be created and implemented in the schools.

## PUBLIC HEALTH

Many Himalayan communities lack access to the basic amenities of modern life—electricity, drinking water, and toilets. This, combined with the prevailing food deficits, sanitation problems, and inadequate health care, contributes to a growing public health crisis in many mountain regions. Nutritional deficiency, maternal disorders, and infectious diseases are common throughout the range. In Nepal, almost half the hill population consumes less than the minimum daily caloric intake, and these people account for 68 percent of the disease burden. Child immunization rates are low, placing children at risk from common childhood diseases, and anemia among young children remains high (in Sikkim, for example, the anemia rate for children younger than three years is 77 percent). The most common diseases and disorders in the Himalaya include diarrhea, iodine deficiency (leading to goiter and cretinism), tuberculosis, leprosy, and various vector-borne diseases such as malaria and encephalitis. The prevalence of HIV and AIDS is not well documented but is believed to be growing, especially in mobile urban societies.

Fortunately, many of the common diseases and disorders are manageable through effective disease intervention, prevention, and curative health services. The Himalayan people

NEPAL: PERCENT OF POPULATION WITH ACCESS TO DRINKING WATER (1985–1999)

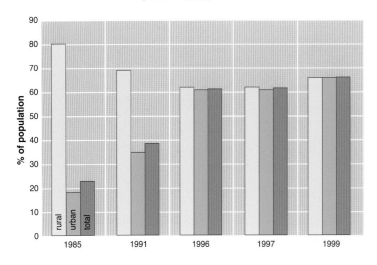

>> Women spend hours waiting at the village spring to fill their household water vessels.

traditionally rely on local health care providers, including faith healers and ayurvedic practitioners. Modern allopathic systems associated with national health services became important across the range during the 1950s. Today, the public health sector in the mountains remains inadequate due to the shortage of trained health care personnel and drug supplies. The shortages are compounded by the fact that much of the Himalayan population is widely dispersed among remote villages in very rugged terrain that is not accessible by motor vehicle. Nonetheless, important successes have been achieved in the area of public health.

Infant mortality rates in the mountain districts are high compared with those of other developing regions, but they are decreasing due to improvements in prenatal care, sanitation, and child immunization. Nepal's infant mortality rate dropped from 172 (per 1,000 live births) in 1971 to 73 in 2000; Bhutan's rate dropped by half in a single decade, from 142 in 1984 to 110

Royal palace of the king of Nepal.

personnel work outside the major metropolitan areas. To overcome these problems, the Himalayan countries have placed an emphasis on training village health workers and community health volunteers, who provide basic health care and share information about disease prevention.

## Governance and Human Rights

The satisfaction of basic human needs, as well as the preservation of the mountain environment, depends on the Himalayan peoples' participation in their own governance. Historically, the region has been controlled by an assortment of tribal coalitions, feudal principalities, monastic orders, or colonial regimes. The establishment of modern nation-states has resulted in new institutional arrangements for organizing the mountain societies. New democratic initiatives are widespread, but in many places, the civil society is threatened with violence. Kashmir, for example, has been embroiled in civil unrest and military action since the independence of India and Pakistan. Separatist activity in Darjeeling and Sikkim, directed toward autonomy, led to civil unrest throughout much of the 1980s. Bhutan evicted tens of thousands of minority Nepalese (the Lotshampas), who now live in refugee camps in eastern Nepal. Tribal agitation in the eastern Himalaya keeps that part of the range mainly off-limits to all but Indian military and government forces. Most recently in Nepal, a Maoist insurrection threatens the peace and security of the country. Under such conditions, it is difficult to achieve an effective local system of governance.

in 2000; and Sikkim's rate dropped from 60 in 1990 to 52 in 1998. Diarrheal diseases, which commonly afflict young children, are prevented by improving access to safe drinking water and enhancing cleanliness and hygiene in homes.

The number of well-staffed and maintained health care facilities is far below that required by the mountain population. This remains a high priority among the Himalayan regions, but the lack of financial resources impairs the delivery of health services. Most people continue to live many hours or even days away from a health clinic of any sort. Few specialized medical

Amid these large controversies are local questions about public administration, gender discrimination, and indigenous and general human rights. The Indian Himalaya operates basically within that country's system of parliamentary democracy. Practically, though, the mountain regions in India have little influence on national political affairs. This underrepresentation was one of the driving forces that led in 2000 to the establishment of the new Indian state of Uttaranchal, located in the Garhwal-Kumaon region (formerly part of the state of Uttar Pradesh). Good governance is absent in Nepal mainly because of the unstable political climate. Nepal has held three elections since reforms in 1990 formed a new parliamentary democracy, but no government has run its full term. The Maoist insurrection that began in 1996 and flared into nationwide violence in 1999 continues to suppress local governance in much of the country. Bhutan is still a monarchy where the king has absolute power, but decentralization of governance was initiated in 1991 with the formation of committees made up of elected village heads (called the Geog Yargye Tshogchung), who are responsible for mobilizing local development. This trend continued in 1998 with the devolution of executive authority in the elected Council of Ministers.

Himalayan women traditionally have greater independence and higher status than do women in the South Asia plains. This is most true among the Tibetan Buddhist cultures and such ethnic groups as the Thakali and Sherpa, where women are major decision makers in the household. Women, of course, also do a great deal of the work in the mountains. The 1992 National Agricultural Survey in Nepal found that women were responsible for up to 71 percent of farm labor and 57 percent of all subsistence work. Yet a 1997 study showed that only 23 percent of women were allowed to dispose of their earnings in capital transactions. The work of women goes largely unrecorded in the economic data and does not necessarily lead to their empowerment in the family. A major form of wealth in the mountains is land. However, land inheritance practices among Himalayan cultures are often prejudiced against women, making them vulnerable to landlessness and poverty.

The absence of women in the nonfarm economy is due in large part to gender inequities in education and training, as well as to the cultural norms that keep women at home. Discriminatory practices against mountain women, who are often unaware of their basic human rights, range from domestic inequity to the trafficking of girls to supply brothels in India and elsewhere.

The cultural norms of the Himalayan societies are generally conservative, and the attitudes and values held by many mountain leaders make it difficult to implement the human rights principles on which the modern Himalayan states are purportedly founded. This applies to women, low-caste groups, ethnic minorities, handicapped people, and other marginal members of society. Nonetheless, human rights laws and legal procedures exist throughout the Himalaya, with varying degrees of implementation and efficacy. The human rights record in Nepal has deteriorated under conditions of the unstable political climate and civil unrest in the country. The governments in the Indian Himalaya maintain that every citizen has the right to social justice, equality, and a decent standard of living. This is difficult to ensure, however, when a large proportion of the mountain population is unaware of these rights or, as in the case of Kashmir, civil unrest and military action overwhelm the human rights cause. Bhutan cautiously places the national sense of peace, happiness, and security above the needs of the individual. This policy is viewed by some as repressive, but a positive outcome is its ability to minimize the corrupting influences of modernity and globalization.

One of the main tasks in the Himalaya is to devise ways of maintaining traditional life and culture and preserving the natural environment while implementing programs needed to reduce poverty and enhance opportunities for human and social development. A balanced development is needed to achieve this goal, one that expands economic opportunities while safeguarding environmental resources and fostering responsible systems of governance. The mountains, despite their imposing appearance, are not immutable. The Himalaya is, in fact, a fragile place, where society and nature are deeply challenged by the new demands of a modern, mobile, and global society.

# PART FOUR  Resources and Conservation

⌃ (part opener) Sangla Valley, Himachal Pradesh.

## » HIMALAYA: LANDSCAPE REGIONS

Land is the single most important resource in the Himalaya, where the vast majority of people are farmers who live in rural villages. The landscape regions of the Himalaya follow a generalized geographic pattern based on elevation. Most Himalayan farmers live in the middle mountains, where elevations range from 1,000 to 5,000 meters. The 2,000-meter zone is the most densely settled area. The outer foothills and plain, with elevations below 1,000 meters, receive a large number of migrants who clear the fertile lowlands and establish new farms.

⌄ HIMALAYA: ELEVATION ZONES

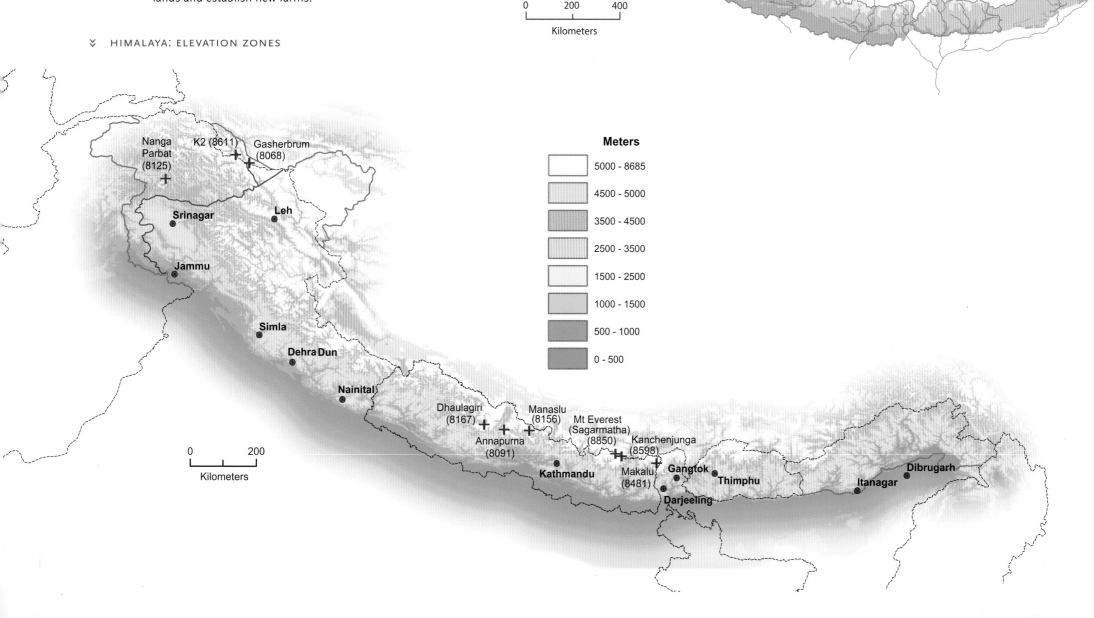

**Legend**

- Tibetan Plateau margins and Great Himalaya (above 5000 meters)
- Hills (1000 to 5000 meters)
- Foothills and tarai (below 1000 meters)
- River

The Himalayan environment provides natural resources for local people as well as for the economic development of mountain states. The villagers traditionally rely on the land for agriculture and livestock grazing; the forests for fuelwood, medicinal herbs, and timber; the rivers and streams for drinking water and crop irrigation; and the native wildlife for game hunting. These age-old practices continue amid the ever-greater requirements of the growing human population. In places where the heightened demand for resources has led to their depletion, the sustainability of village economies and environments may be threatened. Too often the result is human poverty and land degradation. This dilemma has led in recent years to many new and innovative approaches toward resource management and social development in the mountain villages. Such strategies seek to combine the sustainable elements of traditional mountain life with the new opportunities for conservation-based development.

Of particular interest to the Himalayan states is the hydropower capacity of the fast-flowing mountain rivers, the timber value of the forests, and the role of commercial agricultural development, especially horticulture, on the croplands. National policies strive to develop such resources to marshal the forces of economic growth. Unfortunately, such efforts may be managed poorly or inure to the primary benefit of urban localities or plains-based societies. Hence, they may be ineffectual or even contrary to the needs of most mountain people. Moreover, the dual demands of the expanding subsistence and commercial sectors exert a taxing pressure on Himalayan environments and societies. This trend is alarming for many reasons, mainly because it conflicts with human rights and cultural survival, but also because it endangers one of the planet's great biological treasures. Numerous unique and threatened plants and animals are endemic to the Himalaya; if they disappear there, they are gone from the world. In the view of many people, biodiversity is the range's most precious resource, and it must be preserved against the forces of development that threaten to diminish or destroy it.

## Agricultural Land

The majority of Himalayan people are farmers. They till the hillsides and valley bottoms in centuries-old practices to grow grains, vegetables, and pulses, and they graze cattle, sheep, goats, and yaks in the forests and highland pastures. To gain sufficient flat land, the farmers have cut the Himalayan slopes into cascades of terraces, which support crops and distribute irrigation water across the tiny level surfaces. Due to the steepness of the land and the erosive force of the monsoon rain, the loss of topsoil is an ever-present problem on Himalayan farms. By building the terraces, however, the mountain farmers reduce soil erosion by interrupting the length of the slope, and they often plant the strips of intervening land between the terraces with fodder trees and shrubbery to forestall the downslope movement of precious soil. It is only with such careful husbandry that the fragile mountain farms can be maintained over generations of intensive use. The productivity of the land for farming depends on several environmental factors, including the local climate, the soil type, the orientation of the field to the sun (directly influencing energy receipts), and the altitude (affecting temperature), as well as on the application of composted livestock manure or other fer-

The hearth is the nexus of culture and environment for Himalayan households.

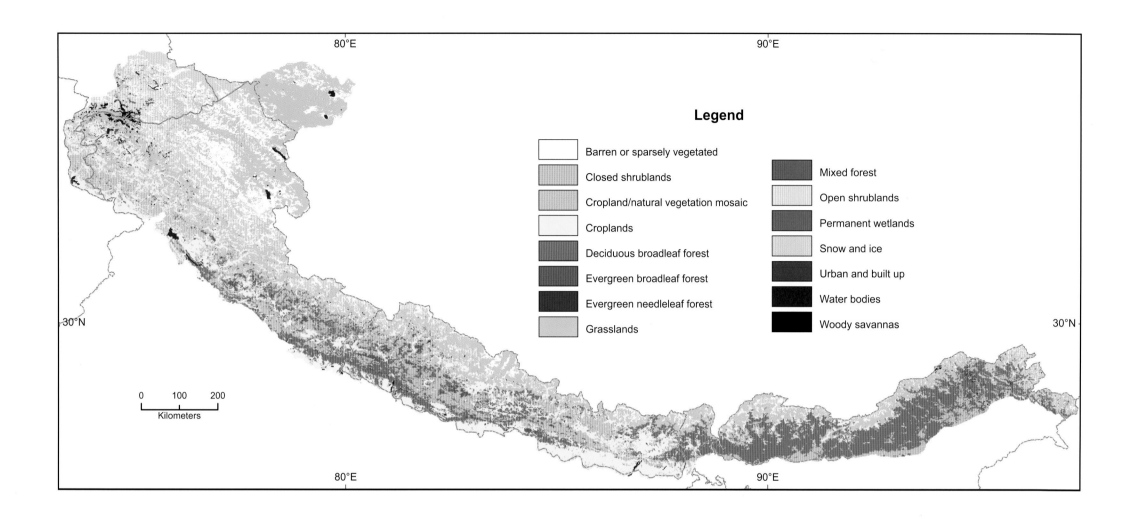

## HIMALAYA: LAND USE

The distribution of land cover types in the Himalaya reflects a combination of environmental and human factors. Soil, climate, and topography influence the natural distribution of plant communities, and human land management has altered the composition of natural land covers. For instance, forests have been cleared and in some cases planted, and grazing lands have expanded into other land categories, where the size of livestock herds has increased. The result is a land cover system that is complex and diverse across the range.

tilizers and associated farming practices. Crop patterns commonly sort themselves according to elevation. Because of the extreme range of altitude in the mountains, the diversity of agricultural zones is great.

The traditional crops grown in the Himalaya include grains such as rice (normally cultivated at lower elevations under irrigation), wheat, millet, and, at higher elevations, where the growing season is short, barley and buckwheat. Corn is grown throughout the middle mountains zone, often interplanted with other grains or with vegetables. Most households maintain home gardens and a few fruit trees. Potatoes were introduced to the Himalaya over a century ago and are important throughout the alpine zones. Village livestock commonly includes cattle, water buffalo, chickens, and goats, which are usually kept around the farmsteads. Sheep are grazed in the more distant highland pastures, often in a seminomadic fashion, and provide villagers with meat as well as wool. A unique type of livestock in the Himalaya is the yak, an oxlike animal indigenous to the high elevations of the range. Yak herds are most commonly found among the Tibetan populations living in the trans-Himalayan zones; they are used as beasts of burden and provide the herders with meat, dairy products, and

102   Illustrated Atlas of the Himalaya

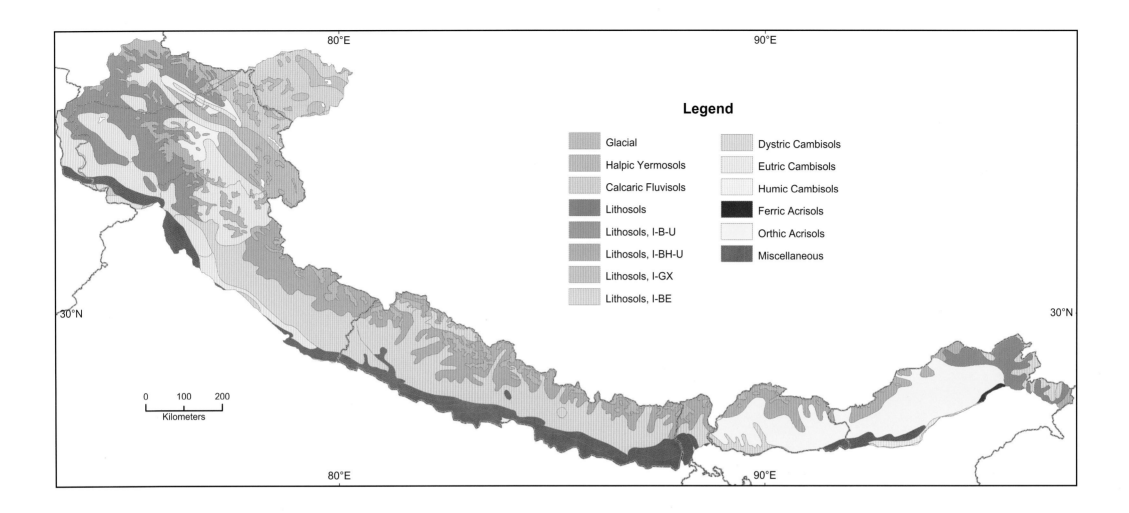

HIMALAYA: SOIL CLASSIFICATION

The agricultural productivity of land is closely tied to soil type. Soil formation is a product of the weathering of parent material, or bedrock, and the addition of organic matter. Meanwhile, heavy rainfall on steep slopes erodes the soil much faster than it is created. Himalayan farmers, in turn, alter the composition and fertility of soils by adding compost and manure. The soil classifications in the map follow the groupings determined by the United Nations Food and Agriculture Organization.

wool. In addition to their farmwork, villagers commonly forage in the forests surrounding their homes for edible tubers, greens, and medicinal herbs. In recent decades, new crops and agricultural practices, including the growing of mushrooms and herbs, beekeeping, dairy processing, and, at higher elevations, fruit orchards such as apple trees, have been introduced in the villages to augment local diets and provide sources of cash income.

Due to the geographic diversity of the Himalaya, as well as to the skewed distribution of people and settlements, the amount of land used for agriculture differs from place to place in the mountains. In the western Himalayan region of Ladakh, where agriculture is possible only under irrigation, less than 2 percent of the land is farmed. Livestock grazing is important in this arid zone, taking full advantage of the seasonal pastures. In Sikkim, the proportion of farmland increases to 10 percent (or closer to 30 percent if we exclude the highest elevations, where agriculture simply is not possible due to year-round cold temperatures). Some of the lowest amounts of land under cultivation are found in the eastern Himalaya. Arunachal Pradesh reports less than 5 percent of its area devoted to farmland. Overall, the amount of agricultural land in the Indian

## HIMALAYA: CROPLAND DISTRIBUTION

The farming systems in the Himalaya reflect environmental conditions as well as cultural systems of adaptation. Shifting cultivation is common in the eastern sector. The low and middle hills of the central and western regions support predominantly mixed grain cultivation. The high mountains are used primarily for livestock grazing.

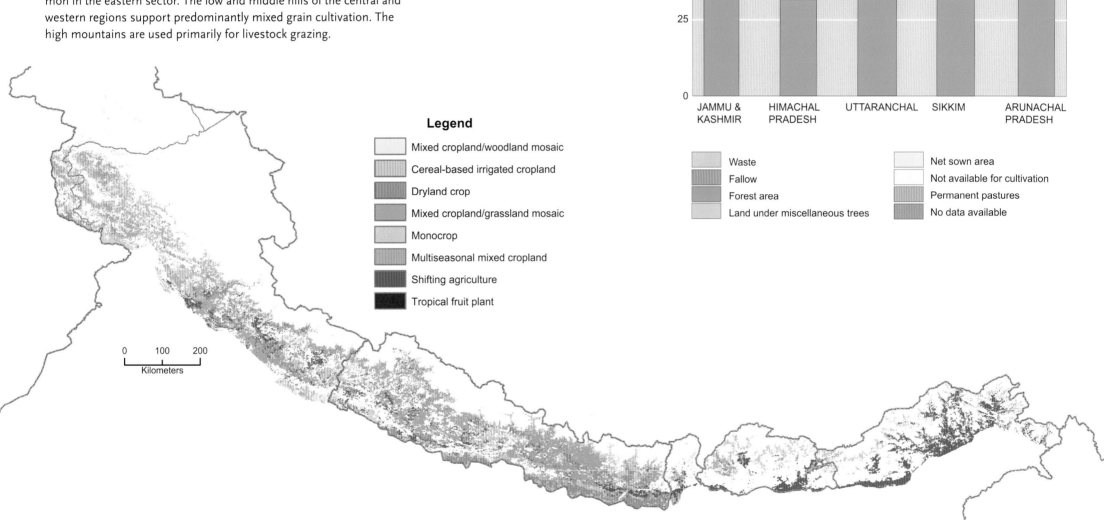

Himalaya is about 10 percent of the total area. Bhutan devotes 16 percent of its land to agriculture of some kind (including pastureland, which accounts for up to 12 percent of the total agricultural holdings), and in Nepal, which exhibits some of the highest population-farmland densities, cropped land alone accounts for more than 17 percent of the country's total area.

With the important exception of Bhutan, where agricultural land is quite evenly distributed and the maximum farm size is limited by law to 10 hectares, the ownership and distribution of Himalayan farmland reflect wealth and caste status. Overall, per capita agricultural landholdings in the mountains are low compared with other farming regions of the world. In Nepal, where farm plots tend to be some of the smallest in the entire range, the average size of landholdings is less than a quarter hectare. The poorest 40 percent of Nepalese farmers own less than 10 percent of the country's total farmland, while the richest 6 percent control over 33 percent of the agricultural area. In Bhutan, almost 50 percent of farmers own less than 1 hectare of land, much less than the government ceiling of 10 hectares. The small size of holdings, the fact that farmland tends to be fragmented, and the low rates of fertilizer use and irrigation combine to present challenging circumstances for

### NEPAL: LAND USE (1999)

| Use | Area (thousand ha) | % of Total |
| --- | --- | --- |
| Cultivated land | 2,968 | 20 |
| Noncultivated land | 998 | 7 |
| Grassland | 1,745 | 12 |
| Forestland | 4,269 | 29 |
| Shrub land/degraded forest | 1,559 | 11 |
| Other uses | 3,179 | 21 |
| *Total* | 14,718 | 100 |

Irrigation canals provide water to crops grown in the arid trans-Himalayan valleys.

Irrigated rice fields in the Marysangdi Valley.

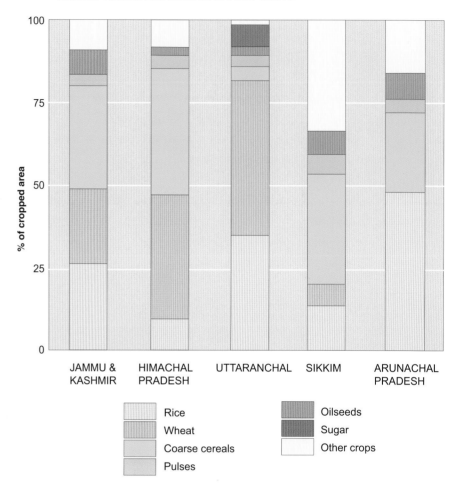

INDIAN HIMALAYA: AGRICULTURAL CROPS

Rice
Wheat
Coarse cereals
Pulses
Oilseeds
Sugar
Other crops

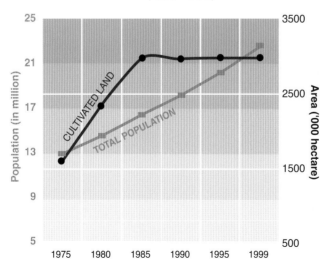

NEPAL: CHANGE IN POPULATION AND CULTIVATED AREA (1975–1999)

poor farmers across the Himalaya. These challenges are compounded by recent agricultural trends, which show that farm holdings are diminishing (per capita farmland throughout the Himalaya declined by 30 percent from 1960 to 1990), and yields are also declining (in Nepal, dropping 15 percent during the 1980–1990 period).

The overall increase in total cultivated area during the past 100 years is due mainly to population growth, which causes people to clear and farm more land, and to migration, whereby people leave overcrowded areas and settle in new places. Some recent increases are due to government schemes and advances

106   Illustrated Atlas of the Himalaya

Shepherds, Nepal.

in agricultural technologies. The archival records kept by the British and later by the Indian governments, as well as those of Nepal, show a steady increase in the cropped area in most localities since 1890, with the most rapid gains observed in the middle mountains and Siwalik foothills. In some places, such as the Garhwal region in the western Himalaya, the annual increase in farmland exceeded 10 percent as early as the beginning of the twentieth century. Between 1950 and 2000, farmland expansion occurred all across the range, with the greatest increases recorded in Nepal's lowland tarai zone. The increase in tarai farmland is the result of the rapid migration of people into the lowlands beginning in the 1960s, when Nepal initiated a malaria eradication program there and introduced a planned resettlement scheme. This policy shifted population density from the crowded hill areas onto the unsettled plains, resulting in significant new pressure on the tarai lands.

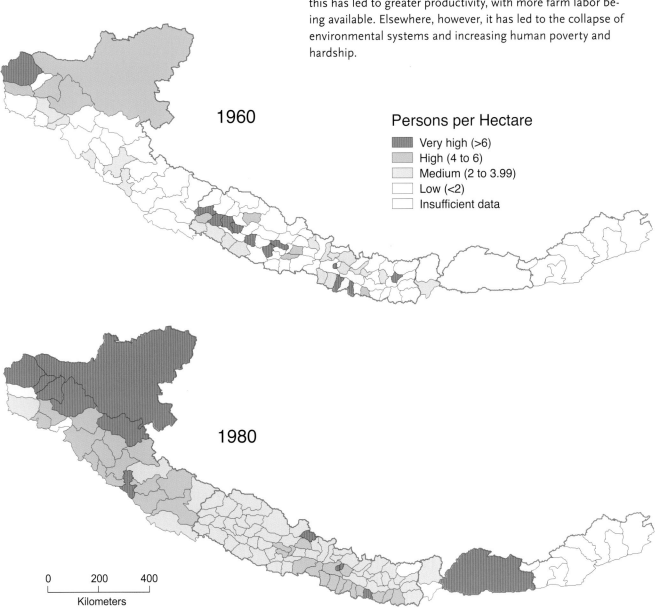

## HIMALAYA: POPULATION AND FARMLAND

With an increasing human population and a finite amount of agricultural land, the number of people who must be supported on the mountain farms is increasing. In some cases, this has led to greater productivity, with more farm labor being available. Elsewhere, however, it has led to the collapse of environmental systems and increasing human poverty and hardship.

In a few localities in the western Himalaya, particularly around the town of Simla and along the lower reaches of the Sutlej River Valley, the amount of land devoted to food grains actually decreased from the 1960s through the 1990s. This was due in part to the extensive development of apple orchards, which displaced traditional farms in some parts of Himachal Pradesh. Other farmland decreases have been recorded in scattered pockets where serious land degradation caused cultivated land to be taken out of production. In such cases, high rates of soil erosion simply rendered the farms infertile.

Some of the sharpest gains in agricultural land during the past three decades occurred in Nepal, where a cluster of fourteen districts in the western region, covering more than 25,000 square kilometers, reported extremely high percentages of farmland growth. This area was historically poverty-stricken, with relatively low population densities, and the increases reflect the new roads, bridges, and irrigation canals that were built in part to support agricultural development. Overall, though, the most significant gains in farmland during the past fifty years have occurred in lowland Nepal, where the annual rates of farmland increase in the tarai districts commonly exceed 10 percent. East of Nepal, in Darjeeling, land records show a recent loss in the amount of farmland devoted to food grains. This is due mainly to the high rates of urbanization in the area and to the expansion of commercial agriculture, mainly tea plantations, at the expense of subsistence fields.

Bhutan has some of the lowest population densities of the Himalayan countries and, consequently, has some of the most favorable natural conditions for agricultural development. This country, however, places great importance on maintaining its forests, and the farming frontier is limited by the kingdom's policy of forest protection. Nevertheless, Bhutan has experienced an expansion of farmland in some formerly unsettled places. The cultivated area in Bhutan increased from 300,000 hectares in 1958 to 554,000 hectares in 1990, a rate of increase that basically matched its population growth. Generally, the expanding farmland in Bhutan is the result of government-

108  Illustrated Atlas of the Himalaya

## HIMALAYA: ANNUAL FARMLAND CHANGE

Land clearing for agriculture is one of the major factors in land cover change in the mountains. Farmers have cleared land for centuries, and much of the agricultural landscape is generations old. During the contemporary period, farmland expansion occurred rapidly until the 1980s, by which time most of the available land conducive to agriculture had already been cleared and converted to farms. There is little land left for additional farmland expansion, and farm yields are likely to increase only by increasing the intensity with which existing farmland is utilized.

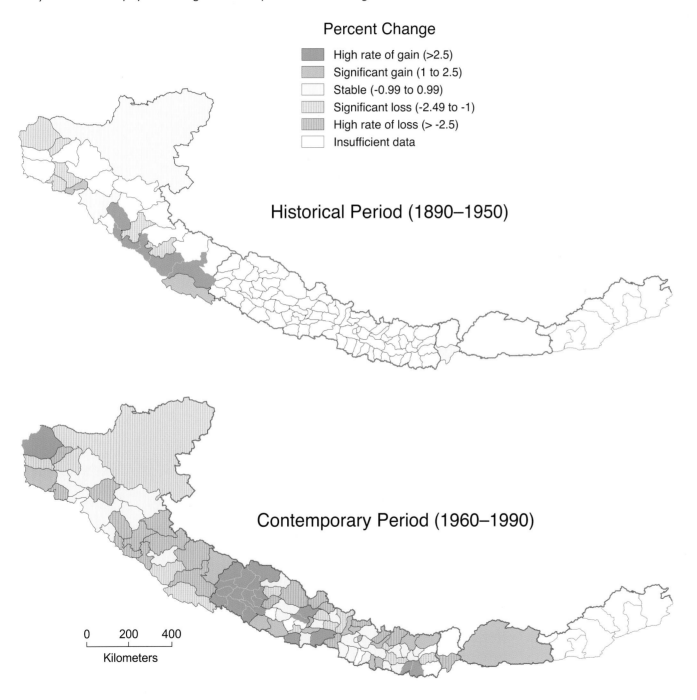

## NEPAL: CULTIVATED AREA

The amount of cultivated land in Nepal relates to population density, climate, and landscape features such as valleys and sloped land.

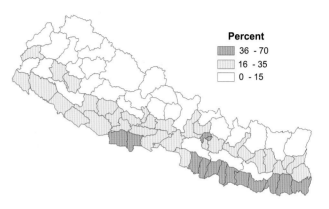

## NEPAL: IRRIGATED AREA

Much of Nepal's agriculture is rain fed; however, rice cultivation generally requires irrigation, for which water is diverted from streams and rivers via both traditional canals and modern technologies.

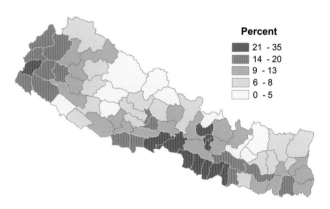

## NEPAL: SLOPING TERRACED AREA

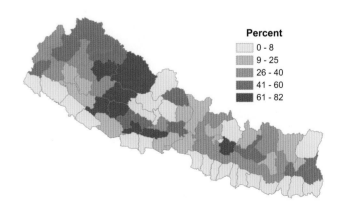

NEPAL: LIVESTOCK AND GRAZING AREA

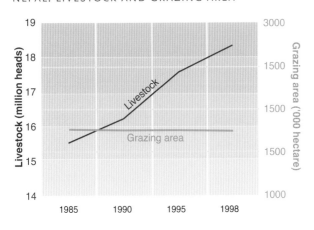

sponsored land settlement schemes and occurs in places where sustainable agricultural growth can be promoted. Some increases, however, reflect spontaneous land clearing by the Nepalese who migrated into the Bhutanese hills during the early decades of the twentieth century.

Only a few densely settled farm areas exist in the northwestern Himalaya. The climate is generally not conducive to agriculture in the trans-Himalayan zone, which occupies a huge area of Ladakh and Zanzkar, but intensive agriculture occurs in scattered localities where irrigation schemes were built.

Low and medium farmland densities were scattered across the lower elevations of the western mountains until the 1970s, when rapid population increases in Kumaon and Garhwal led to the expansion of farmlands throughout the area. The most crowded districts, though, are found in Nepal. Initially, the Nepalese farmers were concentrated in the middle mountains, but since the 1980s, settlement expansion has characterized the tarai zone. The widespread high farming densities in Nepal during both historical times and the present day suggest that land resources have been scarce in that country for quite

NEPAL: GRASSLAND

The main grazing lands occur naturally in the high mountains; in the lowlands, forest has been cleared to create more space for grazing. The high mountain pastures, where villages are located, may extend to altitudes greater than 5,000 meters, making these some of the highest inhabited lands on earth. Yaks and sheep are commonly grazed in the high grasslands. Cattle, goats, and water buffalo are common in the lower pastures.

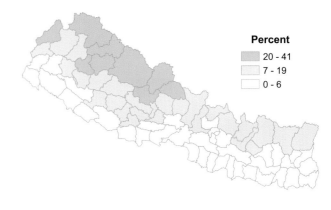

« Monsoon clouds linger over summer grazing lands in the Langtang Valley.

some time. Both Sikkim and Bhutan once enjoyed low pressures on farmland, but the last decades of the twentieth century showed accelerating demands and population densities of up to five persons per hectare of farmland in many localities. The agricultural records in Arunachal Pradesh are difficult to assess, but government estimates show localized land stress in places such as the Ziro Valley and along the fertile stretches of the lower Subansiri and Siang rivers. The eastern region as a whole, though, reports relatively stable agricultural holdings.

### NEPAL: CHEMICAL FERTILIZER USE (1990S)

The use of chemical fertilizers is increasing in many places, especially among larger farms, and results in higher grain yields. Chemical fertilizers lessen the requirements for organic nutrients, which are often scarce, but they add to the problem of soil and water pollution.

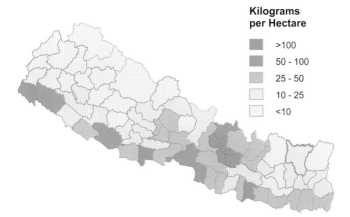

**Kilograms per Hectare**
- >100
- 50 - 100
- 25 - 50
- 10 - 25
- <10

Apple orchard, Himachal Pradesh.

« Woman threshing barley.

# Forests

The Himalaya contains extremely varied forest types, ranging from wet tropical to alpine. The tropical deciduous sal tree *(Shorea robusta)* is a valuable lowland timber species that is widespread in the range. Tropical broadleaf forests also include magnolia, oak, and chestnut, along with locally important pipal *(Ficus religiosa)*. Subtropical pines such as chir pine *(Pinus roxburghii)* occur at intermediate elevations in the middle mountains, giving way at elevations around 2,000 meters to mixed temperate hardwood forests dominated by oak *(Quercus incana)* and rhododendron *(Rhododendron arboreum)*. Temperate forests in the western part of the range include the economically important species *Pinus wallachiana* and *Cupressus torulosa*. Moist subtropical and temperate forests in the central and eastern sectors of the range contain numerous species of bamboo. A variety of pines, spruces, firs, and junipers occurs in the upper temperate and subalpine zones, which converge with the alpine level in a stunted mix of dwarf conifers, willows, and birches.

The condition of the Himalayan forests varies widely across the range, a result of natural geographic factors, such as climate and elevation, but also different degrees of forest use by villagers and commercial loggers. Overall, the Indian Himalaya reports 52 percent of its total area covered by forest, much of which occurs in the sparsely populated eastern region of Arunachal Pradesh (over 90 percent of which is covered by forests). In Nepal, forests account for 29 percent of the total land area (an additional 10 percent of the country is covered in shrubs), and in Bhutan the forested area is currently estimated to be 57 percent of the country's total area (down from 68 percent in the late 1980s). In almost all these areas, however, the forest is decreasing; the estimated overall rate of forest loss is about 1 percent per year across the entire Himalaya. The reported decline is most severe in Nepal, where forests decreased by 24 percent between 1978 and 1994 (currently, the annual deforestation rates in Nepal are 2.3 percent in the hills and 1.3 percent in the tarai); during the same period, however, the area of shrubland increased more than twofold, indicating a trend toward degraded forests rather than their wholesale loss.

The estimates of forest change in the Himalaya point to an overall worsening situation, but the trends are not the same everywhere. Historically, high rates of forest loss occurred in the Garhwal and Kumaon regions of the Indian Himalaya. These averaged about 2 percent per year from 1890 to 1950 and were directly related to timber extractions by the British for purposes of building railroads in the plains and summer resort towns in the hills (for example, Simla and Nainital). The British kept good records of the forest in the western Himalaya, but elsewhere the archives are insufficient to accurately assess forest change during the historical period. However, it is clear that across the range, as farmland increased, forests proportionally decreased.

The land records are more complete for the modern era. Between 1960 and 1990, approximately one-third of all Himalayan districts reported forest loss. Although this trend is alarming, it is not as bad as might be expected from the environmental reports of the 1970s, which suggested that the entire range would be denuded by the early twenty-first century and might become a desert. Much of the contemporary forest loss has occurred in the outer foothills zone, where timber merchants and migrant farmers put considerable pressure on the forests. In Nepal, where the forest conditions are allegedly bad everywhere, some parts of the kingdom are better off than others. The western region of the country, where population densities remain the lowest, maintains a good forest cover, although serious losses are being reported along the Indian border. Rates of forest loss in Bhutan are not readily known, but the overall condition of the forests there remains good. In the far eastern Himalaya, a lack of data makes an accurate assessment of forest change impossible, but the high rates of existing forest cover suggest that little decline has occurred. The exception is in areas where shifting cultivators—practicing a form of agriculture known locally as *jhum*—slash and burn forests

HIMALAYA: FOREST COVER

The exact area of forestland is difficult to discern, but general patterns are calculated based on government data and remote sensing.

**Legend**
- Meadowlands and alpine
- <60% Canopy forest
- >60% Canopy forest
- Cleared land

Hill zone temperate forest.

∨ NEPAL: FOREST CHANGE IN THE TARAI DISTRICTS (1977–1994)

The highest rates of forest clearing occur in the tarai, where land has been settled by migrants and converted to farms and other nonforest uses. These land conversions have led to a geographic pattern whereby most of the large, intact forest areas are now restricted to the steeper slopes of the Siwalik Range or to the national parks and wildlife refuges.

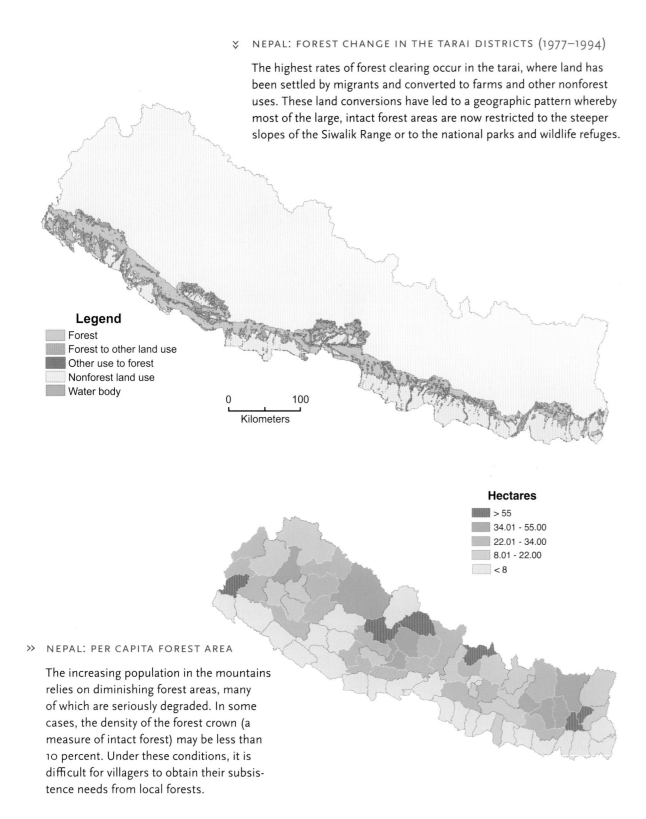

>> NEPAL: PER CAPITA FOREST AREA

The increasing population in the mountains relies on diminishing forest areas, many of which are seriously degraded. In some cases, the density of the forest crown (a measure of intact forest) may be less than 10 percent. Under these conditions, it is difficult for villagers to obtain their subsistence needs from local forests.

Degraded trees heavily lopped for livestock forage in the Dang Valley, tarai.

<< Winter supply of fuelwood.

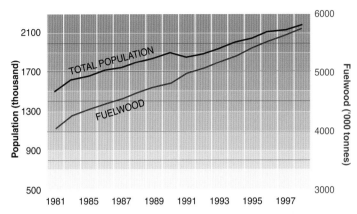

NEPAL: POPULATION GROWTH AND FUELWOOD CONSUMPTION (1981–1997)

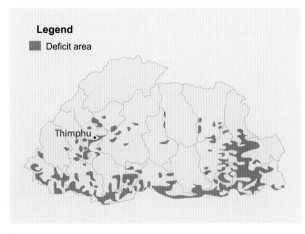

^ BHUTAN: FUELWOOD DEFICIT AREAS

The use of wood for cooking and heating has led to its depletion in heavily populated areas of Bhutan. Efforts to introduce more fuel-efficient stoves, such as ceramic chulla designs, have been effective in some localities in reducing the local fuelwood demand. Overall, though, fuelwood remains the single most important type of energy consumed in the Himalaya.

in an unsustainable fashion, or where timber contractors have illegally purchased logging permits from tribal people and clear forests along roadways and rivers.

The economic and environmental value of forests is relatively easy to ascertain but difficult to measure. They supply fuelwood and fodder for villagers and timber for commercial logging, stabilize slopes, and provide habitat for flora and fauna. Villagers continue to rely on fuelwood for heating and cooking, as well as for the small-scale processing of food, paper, and pottery. On the whole, nearly 80 percent of the energy needs of people living in the Himalaya is met by wood. The villagers continue to graze their livestock in the forests, disturbing the understory growth, and leaf fodder and grass collected from forests are the main sources of household animal feed. In densely populated areas, many of the trees are so heavily lopped for leaf fodder that they no longer maintain their canopy or provide seeds for future generations of trees. The villagers also forage in the forests for food and medicinal plants. These are traditionally sustainable practices, but the high value of some of these nontimber resources compels people to harvest them for commercial purposes. Where local regulations are not enforced, the resulting heavy extractions may take a serious toll on all forest resources.

Even the traditional subsistence needs met by forests, mainly fuelwood and fodder, become problematic when the forest area diminishes greatly. In the western Himalaya, the per capita forest area in the Indus River basin declined from 0.17 hectare in the 1970s to 0.10 hectare in the 1990s. These rates are among the lowest in the entire range. The per capita forest area is a more favorable 0.67 hectare in Garhwal and 0.50 hectare in Kumaon. The forests in the Kulu Valley are in quite good shape, with the per capita forest area currently at about 1 hectare. Nepal, which shows serious rates of forest loss in many districts, also exhibits low per capita forests—0.2 hectare per person overall. Some of the most seriously degraded forests occur in the most densely populated districts of the hill area, resulting in per capita figures of 0.10 hectare in some places. In the tarai, the per capita forest area declined from 0.13 hectare in 1970 to 0.07 hectare in the early 1990s.

The decreasing availability of forests for village use means that people must work harder to obtain the necessary fuelwood and fodder. Greater distances are walked to gather wood for

## HIMALAYA: POPULATION AND FOREST AREA

The subsistence villagers in the Himalaya use forests for fuelwood, livestock fodder, and building material, as well as for medicinal plants and other nonwood resources. Hence, it is important that they have access to sufficient forest areas. In many places, the changes in forest cover, in conjunction with increases in population, have led to a greater number of people relying on dwindling forest resources for their subsistence needs.

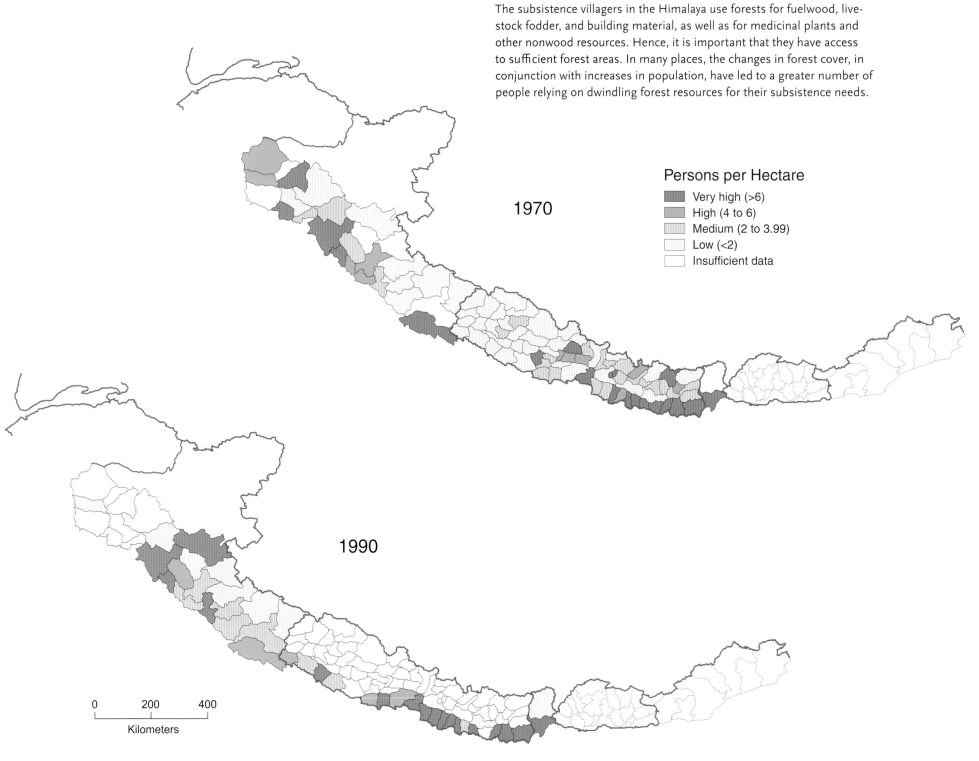

HIMALAYA: FOREST COVER CHANGE

In terms of forests, the historical record is not very good for much of the Himalaya. The exception is the western sector, where the British kept revenue accounts that included forest products. In the contemporary period, government censuses and environmental surveys provide an opportunity to assess the change in forest cover by district. The contemporary period is characterized by significant forest loss, but it is not uniform across the mountains. Some areas register significant losses in forest cover, while other areas report an actual increase in forest area. This patchwork makes it difficult to pinpoint any regional trend.

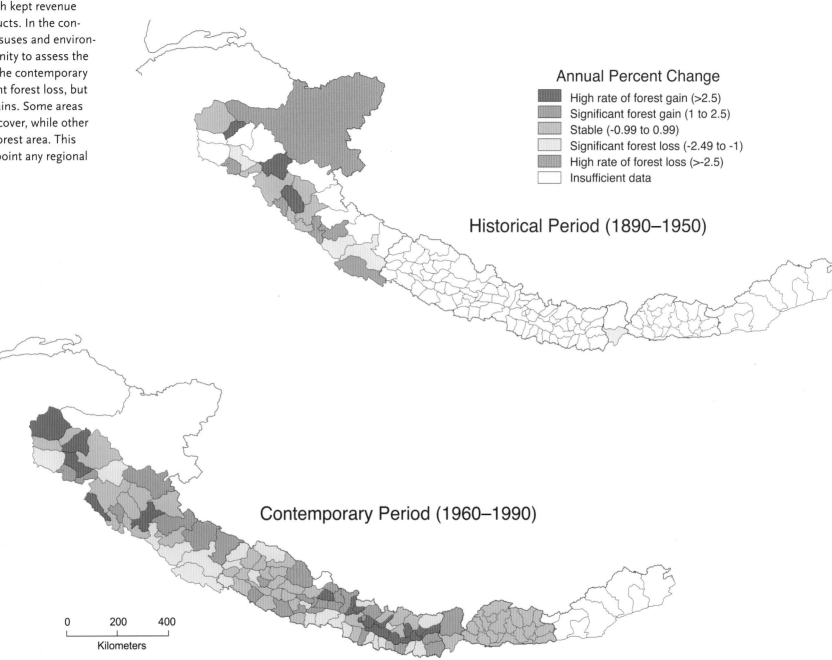

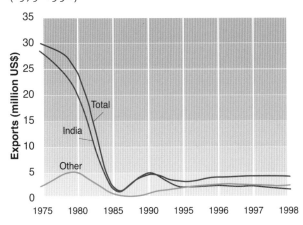

NEPAL: EXPORT OF FOREST PRODUCTS (1975–1998)

fires, often requiring several hours each day. In eastern Nepal, which hosts a large population of refugees from Bhutan, the average time it takes for a household to procure fuelwood has increased from 1.5 hours to 8 hours. Studies in the Kulu Valley, which still has relatively good forest cover, indicate that villagers' continued dependence on wood for cooking and space heating is simply incompatible with maintenance of the forests. Villagers are using more twigs and ground litter to replace scarce fuel logs, with a negative impact on the organic quality of forest soils. Moreover, the soils that are laid bare by the loss of forest canopy and ground cover are overgrazed, making them more prone to erosion and runoff. Under such adverse conditions, forests play a diminished environmental role in absorbing heavy rainfall, storing water, and ameliorating floods. Studies in the Kumaon show that mature forests may retard flood outflow by up to twenty-five days, but under the current stressed conditions of the Almora forests, the outflow is rapid, and local water supplies have consequently decreased.

Commercial logging compounds the subsistence pressures on Himalayan forests. In the absence of tree-planting measures, and where commercial plantations fail to meet local economic or environmental needs, the problem is particularly acute. In the Garhwal region, the exploitation of village forests by timber concessionaires in the 1970s and 1980s led to collective village resistance known as the Chipko movement. The village women in Garhwal sought to restrain the timber cutters by locking their arms around trees, and the tree-hugging motif of the Chipko movement became a worldwide symbol of environmental resistance among indigenous people fighting outsider interests. The Himalayan timber resources are exploited for industrial uses such as paper, construction, packaging, and furniture and for export abroad to generate foreign revenue. Much of the commercial logging is managed by concessionaires, who gain permits, often illegally, from government officials. The timber business provides a lucrative income for both.

In the central Indian Himalaya, the value of logs was calculated to range from 1,980 rupees for chir pine to 8,100 for deodar (*Cedrus deodara*). In Bhutan, the value of wood exports to India increased threefold during the late 1980s. Nepal reported a decline in timber exports between 1975 and 1985, when steps were taken to halt illegal cutting, and formal timber exports to India remained static throughout the 1990s. According to some local forest officials, however, the illegal smuggling of logs across the border actually increased during the same period. Most of the commercial timber cutting in Nepal is limited to the tarai zone, while the logging industry in the Indian Himalaya centers on the pine and cedar forests of the middle mountains. This scenario is likely to change as roads continue to penetrate remote mountain areas everywhere in the range, and timber cutters inevitably follow.

The Himalayan states support a number of new strategies designed to more effectively manage their forest resources. These include efforts to conserve forests by reducing villagers' dependence on trees for heating and cooking energy. Along these lines are more efficient woodstoves, biogas plants that use livestock dung to generate methane gas, and better-insulated homes. A greater emphasis on villager involvement in forest conservation has resulted in the establishment of village-protected forests, forestry user groups, and leaseholder forest management. Such participation, organized through community forestry efforts, establishes more localized and effective regulation of subsistence forest use. In Nepal between 1978 and 1999, the government handed over 0.7 million hectares of state-owned forest to villagers, directly benefiting 6 million people. These forests are now some of the best-managed natural lands in the country.

Commercial timber operations are now more tightly regulated and tied directly to reforestation efforts. The government of Bhutan transferred all commercial logging to the national Department of Forests; the timber is then auctioned to private wood processors. In Nepal and India, commercial logging requires new plantation forests as well as the development of sustainable-yield timber holdings on private lands. A shared feature of all the Himalayan states' efforts toward forest conservation is the establishment of national parks and conservation areas. These designated areas restrict the use of forests for both subsistence and commercial purposes, with the intention of preserving forests for their environmental value. Bhutan currently has almost 1 million hectares of land (20 percent of its total area) under parks and reserves. The area of protected land in Nepal increased from 0.976 million hectares in 1984 to 2.476 million hectares in 1998. It is further proposed that all forested land in the Siwalik foothills zone in Nepal be given nationwide protected status. The Indian Himalaya contains numerous protected areas, totaling over 2 million hectares. Altogether, almost 20 percent of the Himalaya is set aside in various types of designated parks, preserves, and conservation areas that serve multiple purposes, including forest conservation, habitat preservation, and tourism.

Subtropical monsoon forest, central Bhutan.

NEPAL AND BHUTAN: MINERAL DEPOSITS

Although the mineral resources of the Himalaya are significant, a lack of engineering technology and problems of accessibility hamper efforts to develop them.

# Minerals

The bedrock of the Himalaya provides mineral resources that have been exploited on a small scale for many centuries. Lead, zinc, and iron ores, for example, provided materials for weaponry and bridges as early as the fourteenth century. Slate is a traditional roofing material in the local architecture. Early on, copper was wrought into storage vessels and prayer wheels. Systematic mineral exploration, however, began only in the early twentieth century with British and German geologic reconnaissance in the Indian Himalaya. Since independence, the Survey of India and the Wadia Institute of Geology, located in Dehra Dun, have led the mineral surveys in the Indian mountains. Swiss geologists, notably Toni Hagen, began countrywide surveys in Nepal in the 1950s. The 1960s witnessed the beginning of geologic surveys in the kingdom of Bhutan, conducted under the auspices of the Survey of India, with its experienced Himalayan geologists, and Bhutan's Department of Geology and Mines. The main goal behind these various efforts has been to assess the mineral potential that lies beneath the Himalayan surface. With the exception of Bhutan, where less than 30 percent of the country has been mapped geologically in any detail, and the remote and difficult terrain of Arunachal Pradesh, the range has been surveyed quite extensively. The geologic studies indicate important reserves of industrial minerals as well as base metal deposits and gem-quality rocks. The important industrial resources, used for steel and energy production as well as for road-building and construction materials, include dolomite, limestone, gypsum, coal, iron slag, marble, and slate. The base metals include copper ore, and the gemstone-quality minerals include gold, garnet, and tourmaline.

The geologic surveys of India have concentrated much of their exploration efforts in Ladakh and Zanzkar, where iron ore and precious metals are known to exist, and in the lower Siwalik Range near Mussoorie and Dehra Dun, where limestone mining is intensive. The commercial minerals in Nepal include scattered iron ore and copper ore deposits, limestone, and the possibility of rich deposits of precious gemstones. One of the highest mines in the world is located in Nepal; it is more than 5,000 meters above sea level, along the southern flanks of Ganesh Himal, north of the Kathmandu Valley. The mine is purportedly for zinc production, but unsubstantiated reports indicate rich deposits of sapphires, rubies, and other precious stones. The Sikkim Mining Corporation is currently developing mines for the production of copper ore (at the Rangpo Copper Mine), as well as coal and graphite. Bhutan currently mines dolomite, coal, gypsum, and quartzite for export to India and Bangladesh, and limestone for domestic use. Mineral exports contributed only about 1 percent of Bhutan's GDP in the 1990s, but expansion of this sector is clearly part of that country's development strategy. Across the Himalaya, mineral development is geared toward meeting the anticipated needs of domestic industrial growth as well as generating foreign revenue through mineral exports to the southern plains.

## Water Resources

Water is a key resource in the Himalaya, used for drinking and hygiene, irrigation for agriculture, and energy. Its availability at any given place is determined by the local climate, topography, and vegetation, as well as by human management practices. The most important natural sources of water include precipitation, stream flow, and glacial storage. In the outer foothills, groundwater storage in aquifers is also important. The monsoon largely determines the distribution of rainfall, with the greatest amounts occurring across the range in the summer months. Winter storm fronts provide important rainfall mainly in the western regions. The runoff from rain and from melting snow and glaciers shapes the mountains' complex stream systems, whose water flow, in turn, is influenced by the terrain and vegetation. The average annual rate of flow

Tharu fisherfolk on the Rapti River.

in the upper Indus River in the western Himalaya is 115,000 million cubic meters. Rivers originating in the mountains contribute overall about 200,000 million cubic meters to the total flow of the Ganges River as it crosses northern India (10 percent from the Indian Himalaya and 32 percent from Nepal and Tibet). The Brahmaputra River, where it discharges from the Himalaya, has a mean annual flow of 200,000 million cubic meters. An additional 180,000 million cubic meters is added to the Brahmaputra by its tributaries originating in Sikkim, Bhutan, and Arunachal Pradesh, before it empties in the Bay of Bengal. That adds up to a lot of water.

Meltwater from snowfields and glaciers contributes substantially to the discharge of the Himalayan rivers, especially in the arid trans-Himalayan zones and elsewhere during the

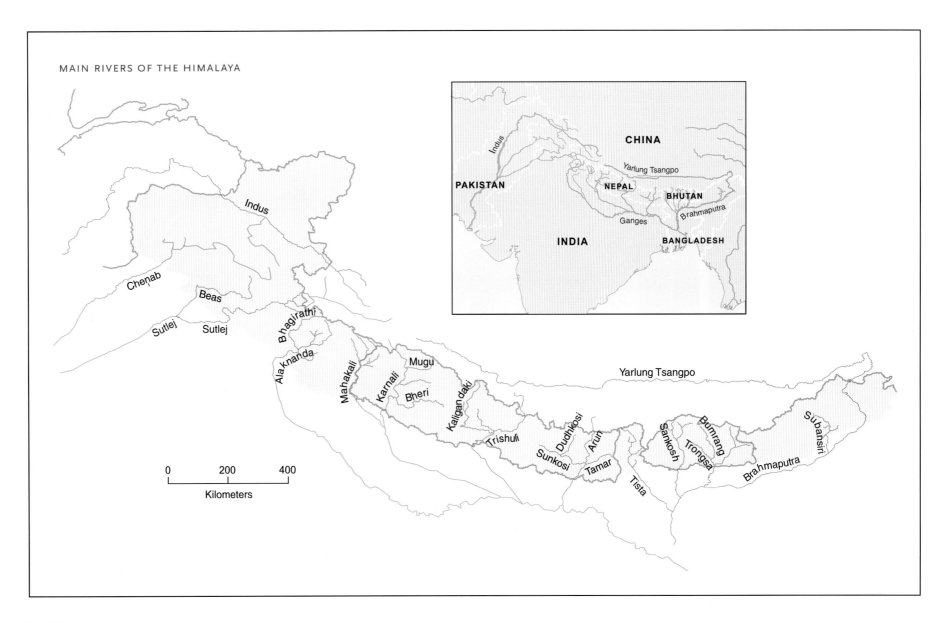

MAIN RIVERS OF THE HIMALAYA

122  Illustrated Atlas of the Himalaya

warm, dry months of the year. Knowledge about the water storage role of glaciers is key to understanding the annual cycles of river flow. Glaciers cover an estimated 15 percent of the entire Hindu Kush–Karakoram–Himalaya mountain belt. An additional 30 to 40 percent of the range is covered seasonally by snow. The percentage of glacier and snow cover diminishes from west to east, with the Indus Mountains containing some of the largest glaciers on earth. These commonly exceed 10 kilometers in length, and several exceed 50 kilometers. By contrast, the glaciers in eastern Nepal, which are that country's largest, rarely exceed 10 kilometers. The diminished glacial area from west to east reflects the more southerly latitudes of the eastern part of the range, as well as topographic and climatic differences.

Three major types of rivers occur in the Himalaya, reflecting their geographic origins. Some originate on the Tibetan Plateau and the northern slope of the High Himalaya and thus predate the geologic thrust of the mountains. These rivers include the Indus in India; the Arun, Kali Gandaki, and Karnali in Nepal; and the Yarlung-Tsangpo (Brahmaputra), whose bend coincides with the eastern syntaxis of the Himalaya near Namche Barwa. They start as tributaries fed principally by melting glaciers and snowfields, and gradually gather volume as they flow southward with the addition of joining streams. A second type of river includes those that originate on the south-facing slopes of the High Himalaya, fed by melting snow and ice as well as by monsoon precipitation. Rivers of this type include the Sutlej and Alaknanda in northwestern India, the Bheri and

>> CHANGING COURSE OF THE KOSI RIVER, NEPAL

Himalayan rivers are not immutable—they may be dammed or rush during floods—nor are they stationary, as this map of the Kosi River demonstrates. Rivers rush out of the mountains carrying heavy loads of sediment; when they reach the plains, they slow and meander, often changing course. This results in flood-prone areas as well as irrigation problems.

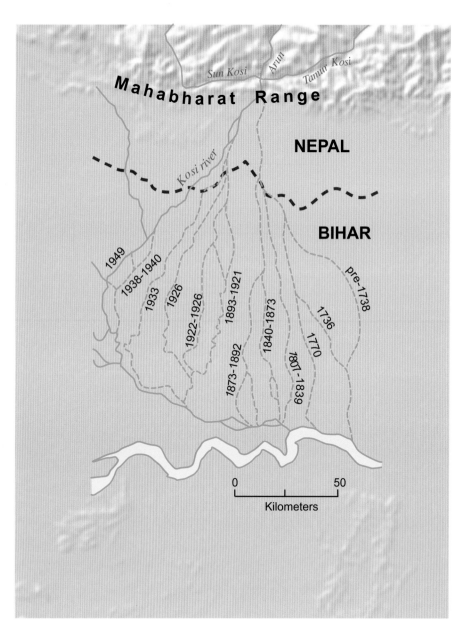

STREAM FLOW IN THREE RIVERS

The discharge of mountain streams shows a pronounced peak in the summer months, when meltwater from snowfields and glaciers is greatest.

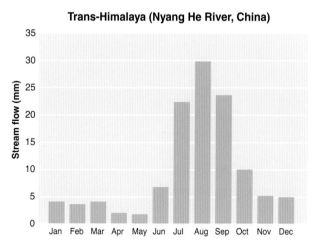

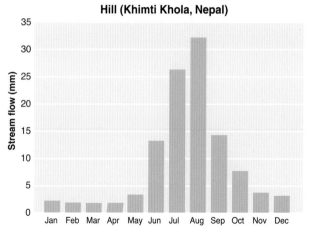

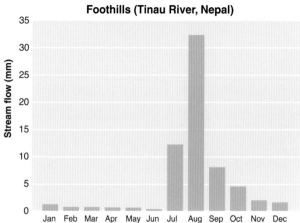

Fish trap laid in a mountain stream.

Hydropower development at the juncture of the Beas and Sutlej rivers is designed to generate electricity for use in Indian cities thousands of kilometers away.

The Kali Gandaki River flows between the Dhaulagiri and Annapurna mountains, carving the deepest gorge in the world.

Marsyangdi in Nepal, the Tista in Sikkim, the Trongsa and Bumrang in Bhutan, and the Subansiri in Arunachal Pradesh. The third type of river has its origins in the middle mountains and outer foothill zones, especially in the Mahabharat Lekh and in the Siwalik Range. Such rivers include the Bagmati, which flows through the Kathmandu Valley, and the Rapti in western Nepal. These rivers generally have smaller and more fluctuating volumes than the rivers that begin in the high mountains, and they have shallow gradients; however, they tend to flood heavily during the monsoon season, when water runs over their banks.

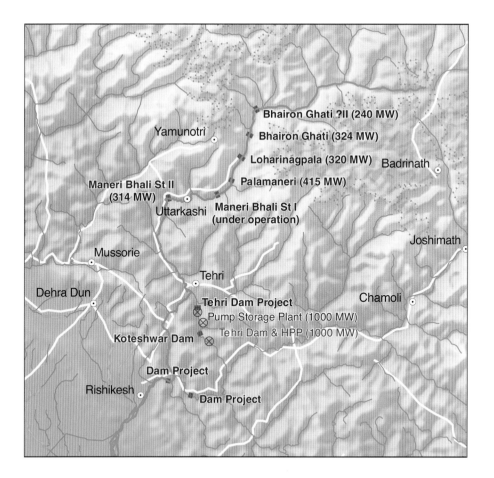

## HYDROELECTRIC PROJECTS IN THE BHAGIRATHI RIVER BASIN

Himalayan rivers are increasingly being developed for their hydropower potential. This may involve small-scale projects designed to meet local village needs or, as in the case of the Bhagirathi River, huge projects that cover entire watersheds and require enormous dams, tunnels, and engineered water diversions. The Tehri Dam project in Uttaranchal has met considerable opposition due to safety concerns about placing the dam in a seismically active region and human rights issues involving the forced relocation of many thousands of people.

Himalayan people have used the rivers and streams for centuries to operate millhouse grinding wheels and to turn waterwheels and prayer flags, thus utilizing the kinetic energy contained in the flow of the water. It was the British, however, who first considered the grand idea of damming the Himalayan waterways to generate massive amounts of electricity. They surveyed the Sutlej River in 1908 with that goal in mind. Nothing was accomplished, however, before India gained its independence in 1947. The country's first prime minister, Jawaharlal Nehru, understood the potential of harnessing water energy in the mountains, but it was not until 1963 that the first major hydroelectric project was built in the Himalaya. The 226-meter-high Bhakra Dam was constructed on the Sutlej River, near the town of Bilaspur, in keeping with the suggestions of the early British surveyors. The Bhakra Dam was designed to generate 1,200 megawatts of electricity for export to the southern industrial plains. Ten years later, in 1974, the Indian government built a second dam on the Beas River, producing 360 megawatts of energy. In the 1990s, a massive hydroelectric construction project linked the Beas and Sutlej rivers in a system of diversion tunnels to produce 660 megawatts of electricity at the Dehar power plant.

The newest and most controversial dam project in the Indian Himalaya is the Tehri Dam, located on the Bhagirathi River in Garhwal. This 260-meter-high dam was first approved by the Indian Planning Commission in 1972 but has met considerable local and international opposition. It is located in an active seismic area, and when the dam is completed, the reservoir will submerge 5,200 hectares of prime farmland and forest and displace more than 100,000 villagers. The opposition has highlighted the potentially devastating impact of earthquakes on the dam and the loss of land rights among indigenous people—issues that are also at the forefront of the general debate about the generation of hydroelectric power in the mountains. These controversies have caused the Tehri Dam project to proceed in fits and starts, and it remains uncompleted.

Nepal, which contains more than 6,000 major rivers and streams, is considered to have great hydropower potential. With 94 percent of Nepal's energy needs still being met by tra-

### INDIAN HIMALAYA: PERCENT OF HOUSEHOLDS WITH ACCESS TO DRINKING WATER AND ELECTRICITY

| State | Drinking Water | | | Electricity | | |
|---|---|---|---|---|---|---|
| | 1981 | 1991 | % Change | 1981 | 1991 | % Change |
| Himachal Pradesh | 44.50 | 77.34 | 73.80 | 54.86 | 87.01 | 58.60 |
| Sikkim | 30.33 | 73.19 | 141.31 | 23.11 | 60.66 | 162.48 |
| Arunachal Pradesh | 43.89 | 70.20 | 59.95 | 15.15 | 40.85 | 169.64 |
| Jammu & Kashmir | 40.28 | NA* | — | 60.87 | NA* | — |

*NA, not available.

Mini-hydropower scheme in the Sangla Valley provides electricity for local villages.

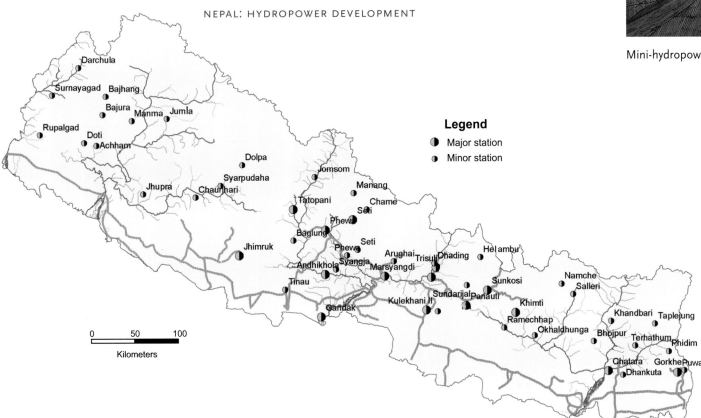

ditional sources such as fuelwood and animal dung, developing its hydroelectric capacity for domestic use is a high priority. Moreover, hydropower is seen as Nepal's major exportable resource of the future. And in one of the world's poorest countries, the huge foreign revenue that could be derived from hydroelectricity is a powerful argument for its development. The primary foreign market would be India, but Nepal theoretically has an energy potential sufficient to meet the needs of more than 700 million South Asians (83,000 megawatts). The current installed capacity is only 252 megawatts, enough to satisfy only a small part of the country's domestic requirement. The proposed large dam and run-of-the-river projects concentrate on the Mahakali and Karnali rivers in the west, the Kali Gandaki in the central part of the country, and the Kosi in the east. The largest of the projects, the 270-meter-high dam at Chisopani on the Karnali River, would generate enough electricity to meet all of Nepal's domestic requirements and still allow some to be exported to India. Additional dams on the

126   Illustrated Atlas of the Himalaya

» GLACIERS, GLACIAL LAKES, AND WATERSHEDS OF NEPAL AND BHUTAN

The glaciers that form among the high peaks of the Himalaya constitute an important source of water for the rivers and streams that make up the mountain watersheds. They eventually converge into the main river systems of South Asia. There is evidence that the Himalayan glaciers are retreating, perhaps caused by global warming. Ridges of glacial till, called moraines, often dam the meltwater from the glaciers to form lakes at their base. Seismic action or other disturbances may cause the dams to burst, unleashing the lake water in devastating floods. These glacial lake outburst floods are a natural hazard of the high mountains.

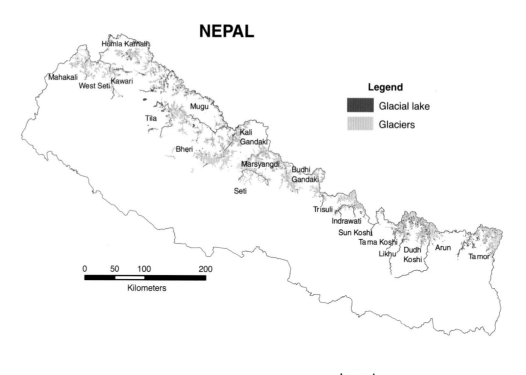

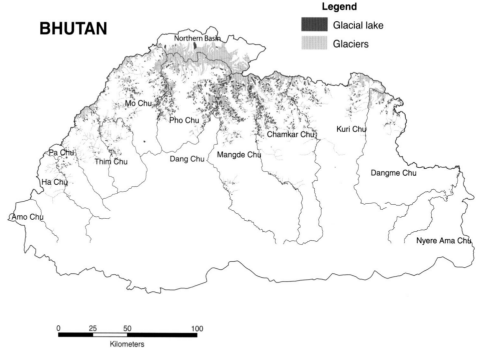

Kosi and Mahakali rivers would generate another 5,400 megawatts, almost all for export to India. But if all the proposed hydropower projects were actually built in Nepal, more than 20 percent of the country's total irrigable land in the hills would be inundated.

Bhutan's energy situation is similar to Nepal's. Fuelwood accounts for 77 percent of total energy consumption in the country, and the energy sector places a high priority on developing its hydroelectric potential. The exploitable hydropower capacity in Bhutan is 6,000 megawatts, sufficient to meet the country's total energy needs, with some left over for export sales. The Chukka Hydro Scheme, which began generating power in 1986, has the capacity to generate 336 megawatts. Bhutan, however, prefers to exploit its hydropower through a combination of large dams and mini- and micro-hydropower plants. The latter, which have up to a 1,000-kilowatt capacity (although most are in the 100-kilowatt range), are considered more viable than large projects because they have lower environmental and social, as well as fiscal, costs. The sustainability of these small hydropower schemes, which are also much easier to implement and maintain, makes them an attractive alternative throughout the Himalaya. In Pakistan, the Agha Khan Foundation has assisted in the installation of 160 small hydropower plants in the Indus Mountains. India set a goal

Resources and Conservation 127

of 600 megawatts of total installed capacity from mini- and micro-hydropower plants by the beginning of the twenty-first century in its Himalayan regions. Nepal installed about 1,000 small hydropower plants during the 1990s (mainly in the 3- to 30-kilowatt range), most of them private ventures. Bhutan's micro-hydropower capacity at the turn of the century included seven small plants with capacities ranging from 300 to 1,000 kilowatts, and twelve micro-schemes with capacities from 10 to 80 kilowatts.

The prospects of small-scale hydropower development for the purposes of local consumption are bright, notwithstanding current technical and infrastructure constraints, and the Himalayan countries are investing heavily in them. But the development of hydropower as an export commodity, which remains very high on the economic priority lists of the Himalayan countries, demands the construction of high dams and run-of-the-river projects, which are both costly and environmentally and socially disruptive. These loom imminent in the future of water resource development across much of the range.

Besides hydropower, the resource potential of the Himalayan waterways applies to other needs such as irrigation, drinking, religious activities, fishing, recreation, and industry. Of these, irrigation is the most important economically. In Nepal, for example, which has a snow-fed river flow estimated at 4,930 cubic meters per second and groundwater reserves in the tarai of about 12 billion cubic meters, irrigation accounts for 90 percent of the total water consumed. The farm area under irrigation in Nepal doubled from 1984 to 1998, increasing from 0.44 million irrigated hectares to 0.88 million hectares. The agricultural development schemes in Nepal, as elsewhere in the range, depend on expanding the role of irrigation. Traditionally, in the arid regions of the Himalaya, water was diverted from small glacial streams through handmade canals and aqueducts to irrigate agricultural terraces in the dry river valleys. Meanwhile, in the rice-growing regions of the southern slopes and lowland valleys, water was traditionally diverted across hand-dug terraces and paddies. The modern irrigation schemes seek to employ pumps, concrete canals, and lock systems using advanced technologies in order to extend cultivated areas and increase yields. The goal is to boost food production to meet rising domestic demand and for export.

Water resources have declined in recent decades in many parts of the Himalaya. Urbanization puts pressure on both groundwater and surface water supplies to meet the expanded and concentrated demand for household and industrial use. The water table in Nepal's Kathmandu Valley has fallen sub-

Glacial lake, Langtang.

&gt;&gt; Gangotri Glacier, Uttaranchal.

stantially in recent decades, resulting in regular water shortages in the city and the need for new water diversion schemes that trap outflow from nearby mountain catchments. Water quality, similarly, has been adversely affected by urbanization, mainly due to contamination from domestic waste and industrial sources. Changes in the land-use patterns in rural areas have affected the water recharge capability of groundwater, as well as the sediment load of rivers and streams. The increased application of chemical fertilizers, pesticides, and herbicides on farm fields, though achieving a desirable increase in yields, has introduced new sources of water contamination. This problem is particularly acute in the heavily commercial agricultural areas, such as orchards.

## Wildlife

The Himalaya covers a huge area that straddles the Palearctic and Indo-Malayan faunal zones and is home to many kinds of animals. In the north and among the high-elevation areas

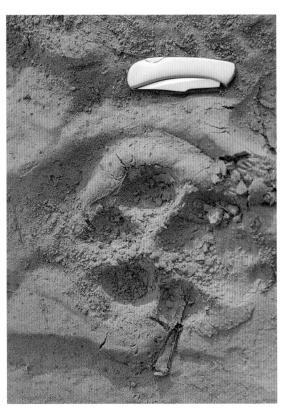

⌃ Pug mark of a royal Bengal tiger.

›› Macaque monkey in a protected forest.

are found blue sheep *(Pseudois nayaur)*, snow leopards, ibex *(Capra ibex sibirica)*, musk deer, red sheep or Ladakh urials *(Ovis orientalis vignei)*, red pandas, and wolves. The temperate zones of the middle mountains host Himalayan black bears, leopards, langurs, wild boars, and barking deer. The southern parts of the range, especially the subtropical foothills, provide habitat for a rich assemblage of wildlife that includes the Asiatic elephant, one-horned rhinoceros, gaur, sloth bear, royal Bengal tiger, gangetic dolphin, and numerous reptiles, birds, and fish. Many of the Himalayan animals are found nowhere else in the world and are globally threatened or endangered species.

Foraging and hunting have always been important components in the subsistence lives of mountain people. Animals traditionally provide meat, hides, and medicines. Many of the new national environmental regulations seek to manage precious wildlife resources by restricting hunting, which often puts government policies in direct conflict with villagers. Game poach-

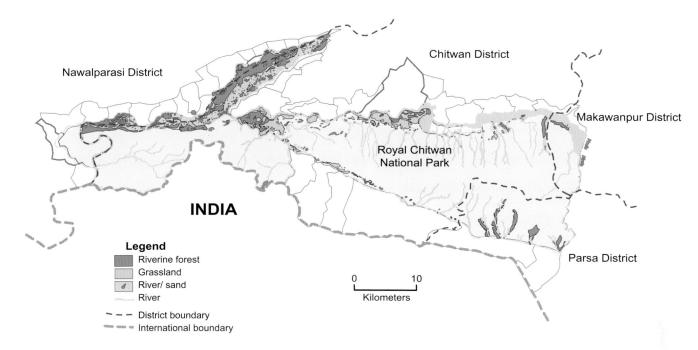

### ROYAL CHITWAN NATIONAL PARK

The oldest park in Nepal, Royal Chitwan National Park in the lowland tarai zone, was established in 1973 primarily as a sanctuary for many endangered animals, especially the Asian one-horned rhino, royal Bengal tiger, Asiatic elephant, gangetic dolphin, and gharial crocodile. Farmers live in the surrounding area. The rhinos' preferred habitat is around the fringes of the park, and the large beasts often enter the farmers' fields, where they do considerable damage to the crops. This has led to serious conflicts between villagers and wildlife. In response, the park service has established buffer zones around the park, where human activity is restricted and wildlife movements are monitored.

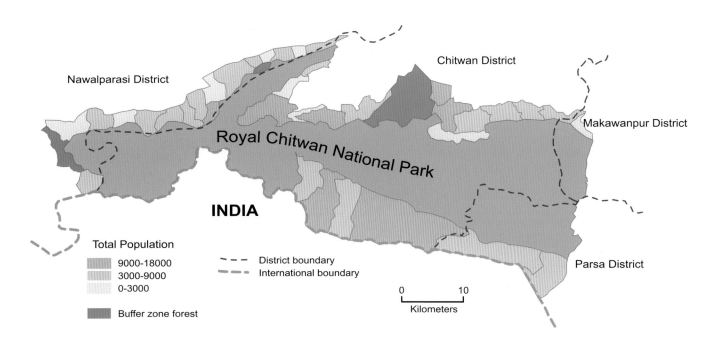

ROYAL CHITWAN NATIONAL PARK AND SURROUNDING VILLAGE POPULATION DENSITY

ing is especially common in and around the national parks, where many rare species with high economic value reside. In some cases, these parks exist mainly on paper, and there are insufficient resources to enforce the rules. Bhutan has banned hunting throughout the country, which affirms its Buddhist orientation, but illegal hunting continues among villagers who seek game for protein and for the illegal sale of pelts and medicines. Hunting is outlawed in most of the national parks in India and Nepal, although both maintain game preserves for regulated hunting purposes. In Nepal, the national parks along the southern border are subject to poaching by both local villagers and hunters coming from across the border in India. Farmers often kill animals that leave the park boundaries out of self-defense or because they are protecting their crops from the marauding wildlife. This problem is particularly acute in cultivated areas around the Chitwan and Royal Bardia national parks, where rhinos and elephants do a great deal of damage to cultivated fields and villages.

The major threat to wildlife resources in the Himalaya, however, comes not from hunting but from habitat destruction. Shifting agriculture and burning encroach on forests in the eastern Himalaya, threatening wildlife habitat in some areas of Arunachal Pradesh. Habitat loss associated with forest clearing, the expansion of farmland, and livestock grazing is a problem throughout Nepal. Its early recognition of this problem prompted the establishment of national parks, which serve to maintain wildlife habitat.

Overgrazing and the enlargement of pastures in the highland regions of northern Bhutan are ongoing problems caused by the 8 percent increase in the domestic livestock population each year. This has transformed the natural habitat of the native blue sheep population. In this case, however, the blue sheep have taken advantage of the expanded grazing land, with a consequent increase in the blue sheep population. In some areas of Bhutan, the problem is now a growing competition between the wild ungulates and domestic livestock. This situation is unique, though, and the dominant theme across much of the Himalaya is the loss of wildlife due to increased poaching and habitat destruction.

## BIOLOGICAL DIVERSITY

The rich biological treasures of the Himalaya may be its most precious natural resource. This unique combination of plant and animal species contributes greatly to the earth's genetic resources, and detailed biological surveys indicate that the mountains constitute one of world's most important centers of biodiversity. In fact, the high percentage of native endangered species in the eastern sector of the range makes it one of the planet's top twenty biodiversity hotspots.

The species diversity throughout the Himalaya is related to its dazzling range of environments and to the convergence of four major biogeographic regions: Palearctic, Indo-Chinese, Indo-Malayan, and Indian subcontinent. The result is a superb concentration of flora and fauna of diverse geographic origins. Additionally, the wide-ranging ecological conditions in the mountains create niche habitats for many endemic species whose worldwide range is restricted to the Himalaya or to small parts of it. Nepal, for example, contains 136 distinct ecosystems ranging from tropical monsoon forests to alpine tundra. The country hosts 35 types of forests, 6,500 species of flowering plants, 656 kinds of butterflies, 844 types of birds, 160 amphibian species, and 181 different mammals. This is an exceedingly large list for such a small country. Moreover, many of Nepal's native species are rare or endangered; hence their presence requires special protection.

The eastern sector of the range, in Sikkim, Bhutan, and Arunachal Pradesh, contains an especially rich assemblage of native plant and animal species. The wet monsoon climate and varied topography ensure the existence of complex ecosystems that permit abundant evolutionary pathways for the development of native species. Tiny, wet Sikkim, for example,

Native rhododendron, Bhutan.

«  Indian great one-horned rhinoceros.

Bharal (blue sheep).

has more than 650 species of orchids. Bhutan lists 47 truly endemic species, but that is just an estimate of the flora and fauna found only in that country. More than 160 species of rare animals have been reported in Bhutan, including the langur, takin, blue sheep, red panda, snow leopard, musk deer, and black-necked crane. Bhutan's strong cultural and ethical commitment to conservation, embedded in the country's Buddhist traditions, has so far permitted the preservation of its rare native species. But the pressures to develop the countryside are strong, and the role of habitat protection is critical to the success of Bhutan's wildlife conservation program.

The eastern regions of Arunachal Pradesh have not yet received the kind of extensive biological surveys needed to assess species diversity and status. But based on limited surveys in India and more extensive ones in adjoining regions of China, a huge number of native plants and animals found only in the eastern Himalaya will require special protection. The historically low human population densities have ensured their survival so far, but with land clearing for agricultural development and the expansion of commercial forestry and other industrial activities, the future status of these rare and endangered species is uncertain. In the meantime, the inventory of species proceeds slowly as the region's scientific and research capacity gradually develops.

In support of conservation efforts, the Himalayan countries have given legal protection to many species of plants and animals. Nepal, for example, protects thirteen plant, twenty-six mammal, nine bird, and three reptile species. But this legal protection is effective only when it is adequately enforced, and unfortunately, this is often not the case. The biodiversity value of domestic plants and animals is often ignored, but Nepal recently took steps to store the germplasm of 8,400 varieties of grains, fruits, vegetables, and agro-horticultural crops, including 680 varieties of rice. As agriculture turns to commercial monoculture, native grains and other crops are threatened with extinction, making such storage increasingly significant. Although a mix of conservation strategies is considered necessary to successfully manage Himalayan biodiversity, one of the most important is to protect habitat, mainly by establishing national and international systems of parks and reserves.

## PARKS AND CONSERVATION AREAS

There are currently about 130 formal protected areas in the Himalaya, covering 13,600 square kilometers. They take the form of national parks, wildlife preserves, conservation areas, and hunting reserves. The western Indian Himalaya contains 23,600 square kilometers of land protected by parks, sanctuaries, and ecological zones, including the oldest conservation landscape in the Himalaya—the Shimla Sanctuary, which was established in 1958. The western Indian parklands include the 4,000 square kilometer Hemis National Park, which protects high desert ecosystems in Ladakh; the 1,413 square kilometer Great Himalaya National Park–Pin Valley National Park in the Kulu and Spiti valleys, home to Himalayan brown bears, musk deer, and snow leopards; and the magnificent Nanda Devi Park, which in 1988 was designated a World Heritage Site by the United Nations. Elsewhere in the Indian Himalaya, the Kanchendzonga National Park was established in 1977 in Sikkim, and the Mouling National Park was set up in the Mishi Hills of Arunachal Pradesh to safeguard some of the biological treasures of the eastern region.

The Himalayan kingdoms of Nepal and Bhutan together contribute twenty-nine conservation areas. About 18 percent of Nepal's total area is under conservation status (including the buffer zones surrounding the national parks), and 20 percent of Bhutan is under formal protection. The oldest national park in Nepal is the 932 square kilometer Royal Chitwan National Park, established in 1973 to protect subtropical habitat in the lowland tarai for populations of endangered tigers, rhinos, crocodiles, and other wildlife. The largest protected area in Nepal is the Annapurna Conservation Area (7,629 square kilometers), which straddles the high mountains in the central part of the country, covering territory ranging from the arid trans-Himalayan valleys to the summit of Annapurna and south to the middle mountains. Local villagers helped design the Annapurna Conservation Area so that it would be compatible with their cultural and economic needs as well as meet its environmental goals. The successful Annapurna project has become a model of sustainable conservation development, and its approach has been adopted worldwide, including in the Makalu Barun, Kanchendzonga, and Manaslu conservation areas, also in Nepal. The most famous national park in Nepal, and quite possibly in the world, is Sagarmatha National Park, which

### HIMALAYAN NATIONAL PARKLANDS

The Himalayan countries have set aside significant areas of protected land in approximately 130 designated areas. Most of these are in the high mountains, and many of the large Himalayan parks have become popular tourist destinations for trekkers and other outdoor enthusiasts. The lowland parks have been established mainly as refuges for rare and endangered wildlife.

Farmers living in the buffer zone of Chitwan National Park sit in makeshift towers at night to guard their fields from marauding rhinos and other wildlife.

Entrance to Royal Chitwan National Park.

Resources and Conservation 135

protects 1,148 square kilometers along the southern flanks of Mount Everest. It is one of the flagship Himalayan parks. With the recent addition of the Makalu Barun Conservation Area to the east and Chomolungma Park in China, situated along the north face of Everest, the world's highest mountain is now protected on all sides.

Bhutan's conservation lands total 9,782 square kilometers, with about 80 percent taken up by the huge Jigme Dorje Wildlife Sanctuary in the northern part of the kingdom. The proposed Black Mountain Park, in the central part of the country, will protect a large area of temperate middle mountains landscape, which receives little conservation attention elsewhere

Villager collects thatch from inside Chitwan National Park.

136   Illustrated Atlas of the Himalaya

in the Himalaya. Much of the remainder of Bhutan's protected land is in the subtropical lowlands in a series of wildlife sanctuaries and forest reserves. The largest of these is the 463 square kilometer Manas Wildlife Sanctuary, once a royal hunting reserve but declared a wildlife sanctuary in 1966. It is the least disturbed protected area in Bhutan's lowland region and is contiguous with the 230 square kilometer Manas Tiger Reserve across the border in India. The southern sanctuaries in Bhutan support important populations of endangered wildlife, and their legal protection in formal conservation areas is the key to maintaining biodiversity in the eastern Himalaya.

The parks and conservation areas in the mountains have

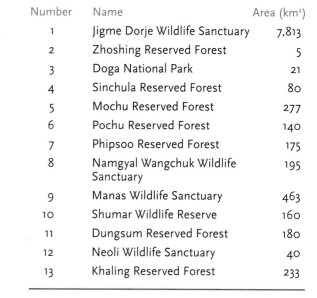

BHUTAN: PROTECTED AREAS

| Number | Name | Area (km²) |
|---|---|---|
| 1 | Jigme Dorje Wildlife Sanctuary | 7,813 |
| 2 | Zhoshing Reserved Forest | 5 |
| 3 | Doga National Park | 21 |
| 4 | Sinchula Reserved Forest | 80 |
| 5 | Mochu Reserved Forest | 277 |
| 6 | Pochu Reserved Forest | 140 |
| 7 | Phipsoo Reserved Forest | 175 |
| 8 | Namgyal Wangchuk Wildlife Sanctuary | 195 |
| 9 | Manas Wildlife Sanctuary | 463 |
| 10 | Shumar Wildlife Reserve | 160 |
| 11 | Dungsum Reserved Forest | 180 |
| 12 | Neoli Wildlife Sanctuary | 40 |
| 13 | Khaling Reserved Forest | 233 |

« NEPAL: PROTECTED AREAS

The largest protected areas in Nepal occupy mountainous terrain and include the conservation areas that adjoin the Annapurna and Makalu summits. Altogether, about 18 percent of Nepal's total land area is under protected status.

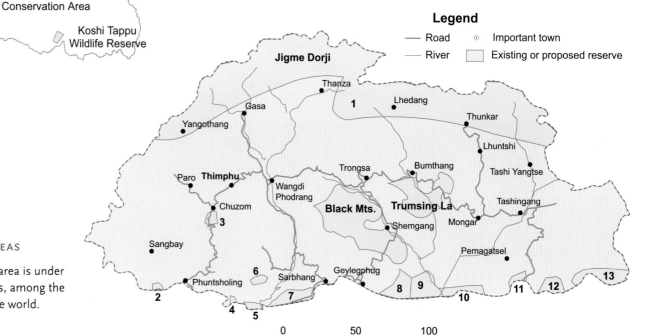

» BHUTAN: PROTECTED AREAS

One-fifth of Bhutan's total area is under designated protected status, among the highest such rankings in the world.

| HIMALAYA: PROTECTED AREA AND BIODIVERSITY | | | | |
|---|---|---|---|---|
| | Bhutan | Nepal | Indian Himalaya | Total Himalaya |
| Total area (km²) | 46,500 | 140,800 | 425,000 | 612,300 |
| Total protected area (km²) | 9,400 | 22,654 | 28,454 | 60,508 |
| No. threatened mammal species | 12 | 17 | 29 | 58 |
| No. threatened bird species | 3 | 2 | 5 | 10 |
| No. threatened reptile species | 1 | 4 | 12 | 17 |
| No. plant species | 5,000 | 6,500 | 15,000 | 26,500 |
| % Endemic flora | 10–15 | 33 | 35 | 32 |
| No. rare or threatened plant species | 5 | 15 | 1,103 | 1,123 |

## ⌃ TARAI ARC CONSERVATION LANDSCAPE

The World Wildlife Fund's Nepal Program, in coordination with governmental agencies in Nepal and India, has developed a model park that straddles the Nepal-India border. The intention is to create a contiguous area of preserved habitat for such large endangered species as the royal Bengal tiger, the one-horned rhino, and the Asian elephant. The transboundary park is a relatively new concept in the Himalaya, and it reflects the fact that the lands required for environmental conservation do not necessarily conform to political boundaries. Such parks require close cooperation among the participating countries.

differing objectives, and some are much more developed than others, but all seek to combine the goals of environmental protection and economic development. This dual purpose recognizes the fact that villagers continue to rely on the environment for natural resources and that their participation in conservation depends on the availability of economic alternatives. Toward this end, the conservation areas have implemented numerous sustainable development programs alongside the environmental regulations. These include agro-forestry and microenterprises such as beekeeping, papermaking, and mushroom cultivation; energy-saving technologies; medicinal plant management; and tourism. Such activities are managed to be compatible with sustainable local resource extractions. The promotion of nature-based and cultural tourism is also an essential component in most of the conservation areas. The revenue generated from visitors seeking to experience both the natural beauty of the mountains and the lifestyles of their resident cultures contributes significantly to both national and village economies. It is believed that such tourism can best be managed in a parkland context where local people are responsible for operating the tourism activities and developing the infrastructure. The premise of the conservation development model is that when local people benefit financially from parks,

they become more deeply involved with them, developing sustainable livelihoods and managing resources wisely and with an outlook toward the future. As a result, conservation areas become more than simply paper parks.

## Future Trends

Geologic forces notwithstanding, the Himalayan peoples have the capacity to manage the future course of society and nature so that both will benefit. History clearly shows that such power lies within the mountain cultures, which possess the wisdom of experience and all the rights of native residency, but future efforts will also need the support of the national governments and of the international community. In the minds of many people, the real challenge is to make decision makers understand that the biological diversity of the Himalaya can be sustained only by maintaining its cultural diversity. This challenge is made more difficult by the impact of globalization, which provides new opportunities for economic growth but threatens the diversity and autonomy of indigenous communities. Reducing population growth and alleviating poverty are fundamental to any formula for achieving a sustainable

Jungle safari in Chitwan National Park.

environmental future. These need to be managed, however, in ways that ensure human rights amid a clean and equitable industry. It is an enormous challenge. Despite the many examples of positive change, the current trends of major social and environmental indicators across the range are in a negative direction.

The worldwide appeal of the Himalaya for its scenic and inspirational value, the magnitude of its societal and environmental problems, and a growing recognition that mountains everywhere play a critical role in the planetary biosphere have led to a new international determination to solve the problems in the Himalaya and safeguard its precious natural environ-

Government-sponsored tree nursery in a central Himalayan village provides seedlings to farmers.

ment. The United Nations Man and Biosphere Programs were among the earliest international efforts to tackle the Himalayan problems. They supported the establishment in 1983 of the International Centre for Integrated Mountain Development (ICIMOD), based in Kathmandu, which is devoted to the sustainable development of the entire Hindu Kush–Karakoram–Himalaya region.

The 1992 United Nations Conference on Environment and Development included Chapter 13, known as the "Mountain Agenda," which promotes the study and protection of mountains around the world, including the important Himalayan region. In the mid-1990s, land cover change in the Himalaya became a component of the International Geosphere-Biosphere Program on Global Change. This inclusion recognized that the environmental health of the Himalaya is tied to that of the entire planet. Such a fundamental recognition led the United Nations to declare 2002 to be the International Year of the Mountain. Such designations, though global in scope, recognize that, at heart, societal and environmental challenges are essentially local ones. They combine with myriad local initiatives in the quest to improve the human condition and to maintain the natural wealth of the Himalaya.

Trekker consults a map along a trail in Langtang National Park.

# PART FIVE Exploration and Travel

## Pilgrimages and Sacred Exploration

The first persons to explore the Himalaya left no written accounts of their travels. What we know about their ancient journeys—as part of the vast migration of humanity across Asia—is sketchy and comes from the incomplete archaeological record. It is only later, with the appearance of outsiders, that we gain a better record of travel and exploration in the mountains. Among the first documented visitors to the Himalaya were spiritual devotees in search of the legendary lands described in ancient Indian and Tibetan religious texts. According to these traditions, the mountains are filled with divine places—remote and hidden from view by snowy peaks, tucked into mysterious, deep valleys—that open the possibility of religious awakening for visitors.

Such spots are safeguarded not only by the harsh geography of the mountains, which makes them difficult to reach, but also by the obscurity of the legends, scriptures, and maps that point to their locations. According to Himalayan custom, the deepest and most holy sanctuaries reveal themselves and give up their spiritual secrets only to those who have acquired the proper religious training and determination. Pilgrims seek them out not because they promise an earthly paradise, such as the Shangri-La of James Hilton's novel *Lost Horizon*, but because they offer a path, literally, to personal salvation. The Tibetans have a word for such places—*beyul*, which signifies

≪ (part opener) Thikse Monastery, Ladakh.

≪ Himalayan temple painting depicts Mount Kailas, a holy peak shared by both Buddhists and Hindus. (Courtesy Hubert Decleer)

a "treasure place." The Hindus call such sacred realms *chetras*, from the Sanskrit word *kshetra*, meaning "holy sphere."

The most famous Tibetan reference to *beyul* is the mythical land of Shambala, which is believed by many to actually exist somewhere deep within the folds of the Himalaya. A promising candidate for Shambala is the mist-wreathed gorge of the Tsangpo-Brahmaputra River, where it wraps around the Namche Barwa massif at the eastern end of the Himalaya before wildly plunging onto the lower plains. The Buddhists know this area as the prophetic land of Pemako, a place "strung with rainbows," where "fortunate ones attain enlightenment." Recent discoveries by the Tibetologist and explorer Ian Baker and his companions show that Pemako's primal landscape is indeed filled with monasteries and pilgrims, botanical wonders, and hidden waterfalls. Other scriptural references to Shambala point to its existence in Sikkim; or as a tributary valley tucked into the flanks of Mount Kanchendzonga; or in Nepal, among the upper watersheds of Helambu, which the Tibetans know as Yolmo; or in the Khumbu landscape inhabited by the Sherpa people; or, finally, in the upper Tang Valley of central Bhutan, where the revered Himalayan saint Padmasambhava lived and meditated.

The sacred Hindu places in the Himalaya center on rivers, especially on their sources amid the springs and glaciers of the high peaks. There is no singular Hindu equivalent of the Buddhist notion of Shambala; rather, Hindus believe that the entire Himalaya is invested with a spiritual quality, which becomes manifest in sacred places called *tirthas*. Like their Buddhist counterparts, Hindu pilgrim-travelers to the sacred mountain places seek not a physical place so much as a metaphorical one. The Indian word *tirtha* also means "truth, forgiveness, and kindness," which suggests that the Hindu concept of sacred place, like the Buddhist one, ultimately rests in the inner, transformational journey of the soul. The ancient Hindu holy text Mahabharata, in the section known as the "Book of the Forest," contains numerous references to India's sacred places in the remote Himalaya. Pilgrims have been seeking those places since time immemorial.

Hindus and Buddhists share deities, religious texts, and spiritual lineages, so it is not surprising that they also share sacred places in the Himalaya. One of the holiest spots for both faiths is Mount Kailas, also known as Mount Meru, the mythical center of the Vedic universe. Mount Kailas, located in the trans-Himalaya, is also the spiritual center of Tibet and

The Bhagirathi River Valley leads to the source of the Ganges above the Gangotri Temple. It is one of the four routes of the "Char Dham" pilgrimage in Uttaranchal.

Exploration and Travel 145

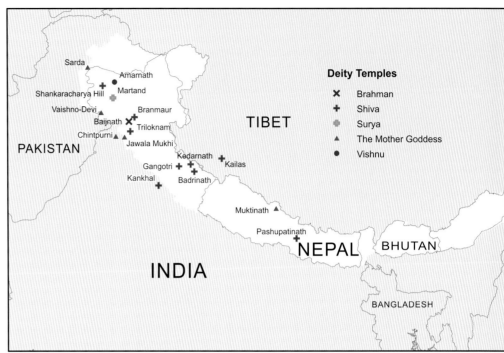

⌃ Thimsigang Monastery, located in the Ladakh Range, sits above a tributary of the Indus River.

» ANCIENT HINDU HOLY PLACES

The ancient scriptures of India refer to numerous holy places in the Himalaya devoted to important deities in the Hindu pantheon. Many of these old sites are still visited by religious pilgrims. The temples located at these places include some of the most revered monuments in the entire Hindu world.

is referred to by Tibetans as Kang Rinpoche, meaning "snow jewel." For Hindus, the phallic shape of Mount Kailas signifies the deity Shiva, who is believed to reign over the entire Himalaya from his abode atop the summit of Kailas; the mountain is thus known as "Shiva's Paradise." The pilgrimage route encircling Mount Kailas is strewn with religious markers of both Hinduism and Buddhism, and devotees of both faiths share the paths and huts as they complete their devotional journeys around the mountain. Buddhists and Hindus also worship together at many of the ancient shrines in the Kathmandu Valley, where the sculptural deities of both groups freely intermingle amid burning incense, fluttering prayer flags, and the sound of bells and chants. The springs and temples at Muktinath, located in the remote Dzong Valley near Nepal's border with Tibet, hold great religious importance for both Hindus and Buddhists, as do lakes Manasarowar in Tibet and Gosainkund in Nepal. The world's tallest mountain, recognized by much of the world as Mount Everest, is known to the Sherpas of Nepal as Sagarmatha, "mother goddess," and to the Tibetans as Chomolungma, "goddess mother of the earth."

Such famous places overshadow the numerous hidden valleys, caves, snow peaks, and river sources scattered across the Himalaya that contain the spirit of Buddhism and Hinduism, as well as that of local tribal religions. The names of these lesser places, known to few outsiders, conjure their importance within the domain of the supernatural world. Sometimes the sacred geography of the Himalaya attains regional proportions. The northern border of Bhutan is dominated by Kangkar Punsum (7,550 meters), which means "white glacier of the three spiritual brothers." Bhutan itself is known in the old texts as either the "lotus garden of the gods" or the "southern garden of paradise." At the foot of the eastern Himalaya is a place called Buxa, meaning "district of all wishes." These allegorical Himalayan toponyms invoke a mythical geography whose sacredness is imbued in the landscape but transcends its physical description. The high peaks, resplendent in their

Old map of Kathmandu.

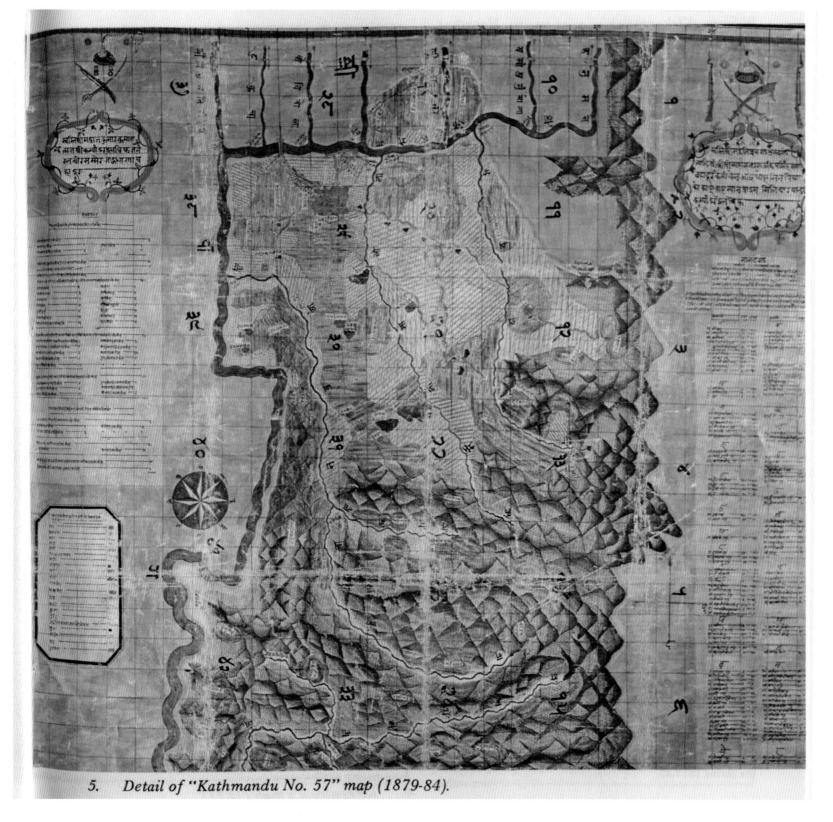

5. Detail of "Kathmandu No. 57" map (1879-84).

GUNGOOTREE, THE SACRED SOURCE OF THE GANGES, 1825.

« Artistic depiction of the source of the Ganges. (Courtesy Mehta Publishers)

» Shivaling Peak (6,543 meters) towers above the Gangotri Glacier at the source of the Ganges River. For Hindu pilgrims, it is the holiest mountain in the Himalaya, its soaring height symbolizing the power of Shiva.

geologic glory, become the thrones of deities. Forests tucked into crevices along the mountains' mighty flanks are filled with demons and ghosts. And the mighty rivers of the Himalaya, fed by alpine snow and glaciers, are considered to be sacred and to sanctify the undulating land through which they flow.

It is these supernatural attributes, and the promise they hold for human salvation, that compel pilgrims to visit the mountains. The arduous nature of the journeys required to reach many of the sites only adds to their religious power. The dangers inherent in the high snowbound passes, precipitous mountain trails, leech-infested forests, and treacherous river crossings are tempered by the practice of meditation and spiritual cleansing at such spots. The Hindu term for pilgrimage, *tirtha-yatra*, requires that the physical act of travel occur simultaneously with religious discipline and devotion. A challenging travel route simply makes the journey all the

Exploration and Travel 149

more worthy of merit. One of the most highly ritualized and demanding pilgrimages is that undertaken by Buddhists who circumambulate Mount Kailas by prostrating themselves along the entire path. This slow movement around the mountain in body lengths, which may take many months, is called *parikrama* and, if successful, is believed to erase the spiritual mistakes of an entire lifetime.

The travel itineraries of the early pilgrims were thus shaped by cosmic designs immanent in the landscape. The maps they relied on to reach their destinations were encoded in the scriptures held closely in their minds. Physical demarcations of the journey, however, were placed in the landscape by native people to mark the pilgrims' progress. Brightly colored prayer flags were hung between trees and across the trail to flutter in the mountain wind, transfixing the stillness of the day. Stone cairns marking the tops of difficult passes were piled high with

« Trongsa Dzong in central Bhutan serves as both an administrative and a religious headquarters for the interior of the kingdom.

˄ Mani wall constructed of carved tablets of scriptural stones lines a walking trail in the central Himalaya.

« The Ganges River flows through Rishikish, an important pilgrimage center, where it descends from the Himalaya and empties onto the Indian plains.

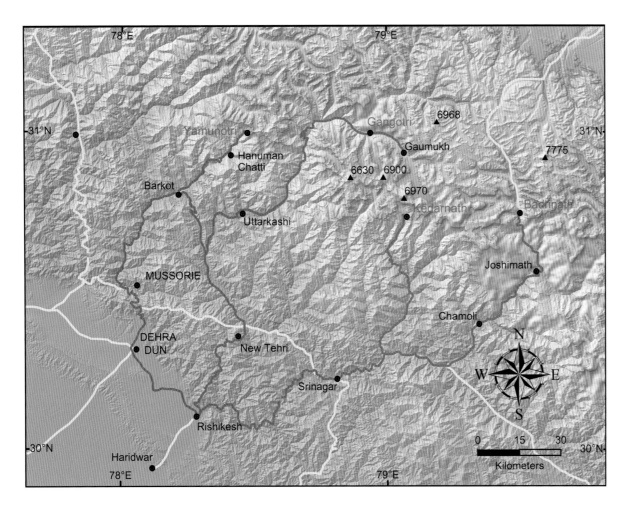

## CHAR DHAM: HINDU PILGRIMAGE TO HIMALAYAN TEMPLES

The Char Dham refers to the pilgrimage to the headwaters of the Ganges River. Four temples are visited by travelers on this sacred route—Yamunotri, Gangotri, Kedarnath, and Badrinath—each corresponding to an important tributary of the Ganges or to another marker associated with the religious landscape. Many Hindus undertake this pilgrimage at some point in their lives, and as a result, the Char Dham sites are visited by thousands of devotees each year.

carved tablets, each scriptural stone placed individually by the hand of a traveler. Pyramid-shaped stupas containing the relics of saints straddled the pathways. Traditional resting spots known as *dharamasalas* or *chautaras*, often marked by the presence of a sacred ficus tree, provided the pilgrim-travelers with places to sleep and eat along the way. Holy springs and temples situated along many of the routes reminded the pilgrims of the purpose of their journeys. Outstanding geologic features in the landscape were explained as acts of God or of saints.

In the minds of the faithful, the entire region is sanctified by the presence of holy mountains such as Kailas, Nanda Devi, and Machapuchhare, whose sacred summits stand as sentinels in a cosmic counterbalance to the valleys formed by the Himalayan rivers. Nanga Parbat and Namche Barwa appear at opposite ends of the range, acting as its topographic anchors, and the arid landscapes of the upper Indus River juxtapose the wet lushness in the eastern bend of the Brahmaputra. The intervening space contains all the possibilities of the human spirit and, consequently, of religious travel.

The greatest journeys of the pilgrims transcend the geographic markers on the land, relying on them merely as portals into the mysterious supernatural realms. The cliff faces and waterfalls, deep mist-shrouded valleys, caves, and summits, juggernauts in the physical world, are metaphorical doorways into the hidden sacred worlds. The act of travel transforms the pilgrims' journeys through these physical landscapes into inner quests for spiritual understanding and enlightenment. It is *this* potential of religious travel that has lured people into some of the harshest and most dangerous terrain in the world.

The early pilgrimage routes are described in the religious texts with varying degrees of certainty. The epic Mahabharata (circa 300 BC) provides an account of the Grand Pilgrimage for Hindus, which encompasses all of India. The circuit of the Grand Pilgrimage includes the Himalayan destinations of

« Hindu devotees in religious worship along the banks of the Ganges River.

Yoni, Varah (Baramula), and Vadava (where the Yamuna and Ganges rivers originate) in the western range, and Salgrama and Gauriskara in the central range in what is now Nepal. The direction of travel to those places is clockwise along prescribed routes, which over the ages have become busy corridors of pilgrim travel. Many of the religious destinations known in the Mahabharata are still popular pilgrimage sites, and in the more than 2,000 years that have lapsed since the writing of that great epic, numerous holy places have been added to the Himalayan pilgrimage map. Kashmir contains the sacred places of Amarnath and Shankaracharya Hill. The headwaters region of the Ganges River is populated by numerous shrines and pilgrim towns, including the great temples at Gangotri, Kedarnath, Jamnotri, and Badrinath. Two of Hinduism's most revered places of pilgrimage are located in Nepal—the Shiva temple complex at Pashupatinath in the Kathmandu Valley, and Muktinath, located north of the Annapurna massif, which is a famous temple complex dedicated to the deity Vishnu.

Muktinath is known in the Mahabharata as Saligrama, denoting the presence of the sacred black rocks known as saligrams. These smooth, polished stones are found embedded in river deposits and display the spiraling pattern of fossilized ammonites. The wheel-like design is regarded by Hindus as a chakra and has sacred ritual importance. Its physical setting high in the mountains at the head of a river gives Muktinath an auspicious location. The site contains a series of springs diverted into 108 fountains, a natural gas fire located in a small cave, a large number of very old Hindu temples, ritual bathing areas, and a sacred grove of ancient poplar trees. The convergence of geographic factors and the natural elements of fire, water, and fossils gives the place its divine presence as Muktichhetra, or "salvation place."

Tibetan Buddhists also revere Muktinath as a pilgrimage destination. They call it Chumig Gyatsa, meaning the place of "a hundred odd springs." For Tibetans, Muktinath is the residence of the bodhisattva Avalokiteswara (the Buddhist avatar of Vishnu), and it was visited more than 1,200 years ago by the famous Buddhist saint Padmasambhava, who first introduced Buddhism to Tibet. Tibetans believe that the trees in the poplar grove come from the walking sticks of eighty-four ancient Indian Buddhist magicians. They consider the saligrams to be representations of the Tibetan serpent deity known as Gawo Jagpa. And they venerate the natural gas fires as the sexual union of tantric male and female deities. Alongside the Hindu temples are Buddhist monasteries filled with Tibetan icons, and Buddhist and Hindu pilgrims share the trails into Muktinath.

The scale of the Tibetans' sacred geography spirals outward from the village into much larger regional entities. The smallest boundaries of sanctified space enclose individual homes, which require ritual purification to remain sacred; these boundaries are metaphorically drawn around the perimeter of villages as *chos-skor*, or "scripture circuits." When tainted houses are ritually purified by the prayers of lamas, the unclean spirits must be carried beyond the sacred boundary of the village before they can be discarded. The sacred places, which are guarded by female agents known as *dakinis*, are enumerated in the ancient Tibetan Buddhist texts and include large parts of the countries that adjoin the Tibetan Plateau. Muktinath and other places in Nepal (Mugu, Mustang, and Dolpo) are within these boundaries, as are many places in China (Yunnan and Amdo) and India (Kashmir, Ladakh, and Zanzkar).

Like the Hindu circuits, the routes to the Tibetan pilgrimage destinations are described in the ancient texts, many of which remain inscrutable to outsiders because of their reference to esoteric rituals and mythological places. They are also drawn in maps that are more pictograph than cartography, which is appropriate, given the fact that they chart a course through a sa-

Prayer flags line a Himalayan trail.

Exploration and Travel 153

☆ Pagoda-style temple, Kathmandu.

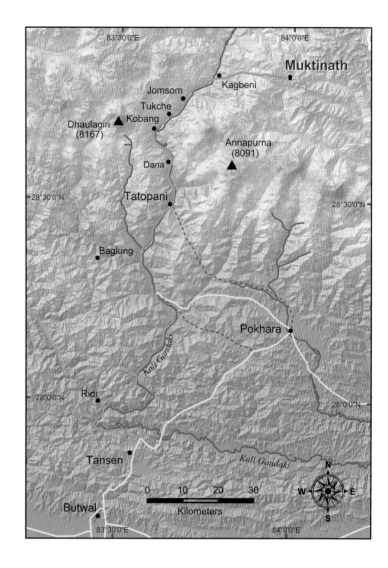

>> PILGRIMAGE ROUTE TO MUKTINATH

The pilgrims' route to Muktinath takes them north from the Indian plain through the rugged middle mountains of Nepal and into the Kali Gandaki Gorge—the world's deepest canyon. The gorge leads the pilgrims to the Dzong Valley, located on the north side of Annapurna. Here, the sacred springs and temples of Muktinath are found amid the soaring peaks and barren slopes of Mustang.

cred topography that exists mainly in the Buddhist mind. The religious maps and pilgrim itineraries conjure a magical world in which a divine geography is encrypted in the obtuse genealogies of deities and in the symbolic spiraling forms called mandalas, captured geologically in the saligrams.

The graphic mandalas are representations of a sacred landscape and denote a progression, or "stages of enlightenment," through which pilgrims metaphorically journey on the road to salvation. The mandala design appears often in the mountains as brightly painted frescoes in monasteries, devotional wall hangings called *thankas*, or inscriptions in stone along the paths. They serve as offerings to the gods and as wayfinding guides for the pilgrims and other spiritual travelers. The mandala design is repeated in the religious buildings of both Hindus and Buddhists, whose spatial orientations and architecture mimic a grand cosmic design symbolized by the mountain terrain itself. The pagoda-style buildings and the spiraling stupas—religious structures found throughout the Himalaya—exist as pure symbolic forms to portray the cosmic elements of the mandala, emanating in their ordination and shape a highly stylized rendition of sanctified space. This aspect of Himalayan architecture is especially striking in the complex and multistoried *caitya* structures of the Kathmandu Valley. These complex religious monuments, found through-

Yak caravan from Tibet passes beneath religious paintings along an ancient Himalayan trading route in the Khumbu region of Nepal.

out Nepal's central valley, contain antediluvian inscriptions and are designed and precisely placed to represent the Kathmandu Valley as the center of the world. Pilgrim-travelers pass through this maze of monuments as if following a neatly drawn map, on a route that replicates in miniature the great spiraling mandala of the entire Himalaya. The circuitous routes of the Hindu and Buddhist pilgrims, writ small in the circumambulations of devotees at specific religious monuments, or writ large in the wanderings of mendicants across the range in search of hidden sanctuaries, unveil the native geography of the Himalaya.

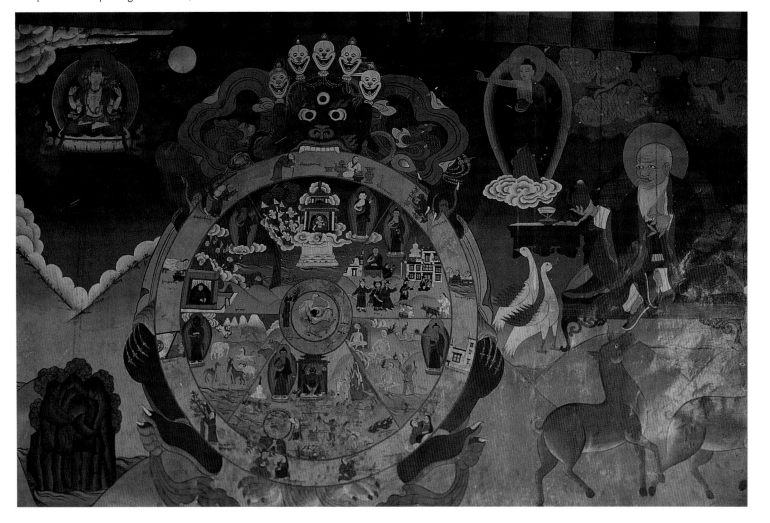

Temple mural depicting a mandala, or sacred wheel of life.

## Early Foreign Exploration

With the exception of the Jesuit priests, whose religious convictions urged them forward into the most remote and inhospitable outposts of the Himalaya, the travelers who followed the pilgrims' paths into the mountains were compelled by reasons other than spiritual practice. Some were military adventurers bent on conquering new lands, others were itinerant merchants working the bazaars and caravansaries located along ancient trade routes, and more than a few were swashbuckling loners compelled by a personal quest for geographic knowledge. Regardless of their origin and disposition, the earliest foreigners entering the Himalayan region shared a fundamental quality: none possessed accurate information about the mountains.

The Europeans first came to know about the mountains of the Indian subcontinent when the army of Alexander the Great crossed the Hindu Kush Range in about 325 BC. Local traditions place his troops in the Hunza and Swat valleys of present-day Pakistan, and perhaps even as far as Kashmir, although the official records do not verify this. Notwithstanding the debate over its easternmost penetration, Alexander's march into the Orient extended the Europeans' map of the world by more than 3,000 kilometers. It uncovered nothing of substance, however, about the interior of the Himalaya. Many centuries passed before accounts appeared that filled in some of the blank spaces of highland Asia. The Chinese explorer Hsuan Tsang undertook in AD 629 a formidable journey to India that skirted the harsh northern ranges and high steppes of Tibet before crossing the Tien Shan mountains and picking up the well-established trade route south over the Hindu Kush and onto the southern plains of Afghanistan. His account of the

Central Asian highlands is filled with descriptions of treacherous trails, deep snow, and penetrating cold, but it provides little actual geographic information about the mountains.

Altogether, Hsuan spent over a decade exploring the subcontinent, which he described quite well, but unfortunately, he only barely set foot into the Himalaya and returned to China along the known route through Afghanistan to Samarkand, where his Indian stories intermingled with those of countless traders and caravans on the ancient Silk Road. Hsuan remarked in great detail on the Mongol herdsmen living among the steppes north of the Himalaya, who in the thirteenth century would gather under Genghis Khan as the Mongol horde. The thundering Mongol horsemen crisscrossed the northern flanks of the Himalaya as they plundered China to the east and the Muslim civilizations in the west, effectively keeping all prospective travelers at bay. It was only with the waning of the Mongol empire that foreigners once again visited the mountains of Central Asia. One of the first and most famous of these travelers was the Venetian trader Marco Polo.

By some accounts, Polo never made it to China at all, instead fabricating his wondrous tales of the mysterious East from stories he had heard from other travelers while he maintained a residence in Persia. Even his most ardent backers admit that it is impossible to follow Polo's "footsteps" east of Persia because his diary, which was published as *Description of the World*, contains no true itinerary of his travels but rather provides "geographic snatches" from various locales. Nonetheless, from Polo's descriptions of his four-year journey east to China, it is clear that, once again, the Himalaya proper was skirted in favor of the easier routes north or south of the range. He described in minor detail his crossing of the Hindu Kush and Pamir ranges in AD 1273 and his descent of the steppes of western China before joining the Silk Road, and he recounted traveling from oasis to oasis along the fringe of the lifeless Takla Makan desert, but Polo provided little account of the great mountains immediately south of his route.

The first reliable accounts of the Himalaya did not appear until some 300 years after Marco Polo returned to Venice, and they were provided not by military conquerors or traders, for whom geography plays such a defining role, but by Jesuit priests from Portugal. Like the Hindu and Buddhist pilgrims, the Jesuits were in search of religious mysteries. Unlike their predecessors, however, they sought not personal salvation, which was more easily attainable in the cathedrals of Europe, but rather the confirmation of rumors about pockets of Nestorian Christians believed to be living secretly somewhere high in the remote Himalaya. If the rumors were true, such settlements were certainly deemed worthy of clerical investigation. Some of the priests sought converts among the native people, imagining a Christian kingdom in the mountains alongside those of the Buddhists and Hindus.

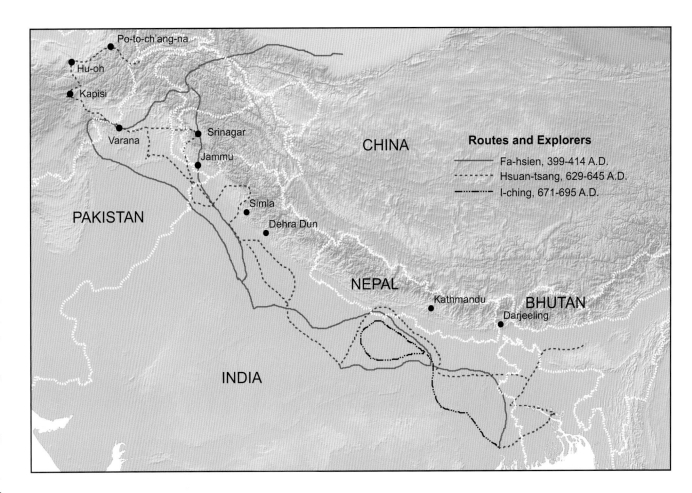

EXPLORATIONS BY CHINESE TRAVELERS

The first persons to explore the Himalaya were Chinese travelers en route to India. Their journeys into the mountains were restricted to the western region, and only occasionally did they veer from the established trade routes. The journals of these Chinese explorers provide some of the original firsthand accounts of Himalayan geography and culture.

Exploration and Travel 157

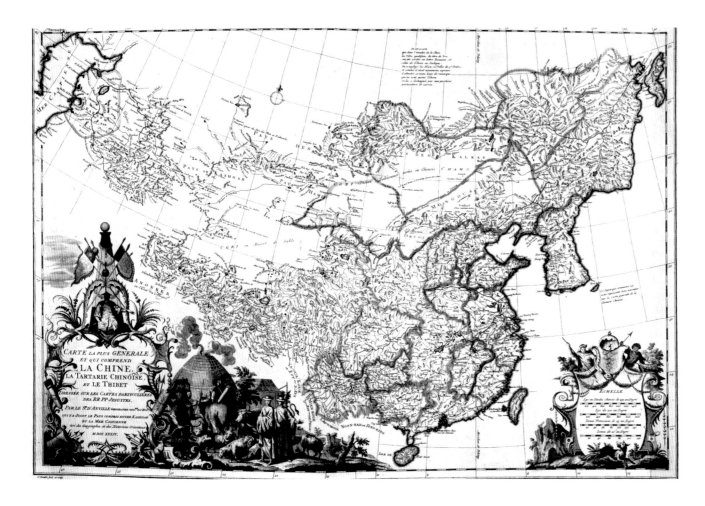

D'ANVILLE MAP

The Himalaya made its cartographic debut on Jean Baptiste D'Anville's famous 1737 map of China. Prior to this, the mountains failed to show up on Western maps of the world or were limited to minor details that failed to convey their impressive topographic presence in Asia. (Courtesy of Library of Congress)

The earliest cartographic descriptions of the Himalayan range that became known to Europeans were made by the Jesuit priests, including a map drawn in 1590 by Father Anthony Monserrate. Monserrate never actually penetrated the Himalaya, but his map was based on the accounts of others who had, and it stands as probably the first European map of the mountains that bears any resemblance to reality. Firsthand accounts of the Jesuits' journeys into the Himalaya appeared a quarter of a century later, in 1625, with the travels of Father Antonio Andrade. He left his mission in Goa on the Indian coast and journeyed into the western part of the range in the guise of a Hindu pilgrim, undertaking what is most likely the first proper exploration of the Himalaya by a Westerner. Andrade's route took him through Kumaon and up the Alaknanda River to the Hindu pilgrimage destination at Badrinath, near the source of the Ganges River. He described crossing the upper reaches of the Vishnu-Ganga in his letters to superiors: "Not as hitherto by difficult rope-bridges, but over bridges formed by frozen masses of snow, which fill up the whole width of the river and underneath which the river-water breaks itself a foaming passage." Andrade lingered for some days among the Hindu temples in the region, noting the many pilgrims who visited them and describing some of their esoteric practices of faith. He eventually proceeded as far as the Mana La, a 5,608-meter pass overlooking the Tibetan Plateau, but contrary to accounts that have him traveling all the way to Lake Manasarowar, Andrade never made it beyond the high pass; he was turned back by deep snow and the death threats of a local warlord.

Andrade's journey precipitated speculation about the source of the Ganges River. He had described a small glacial lake near the headwaters of the Vishnu-Ganga, which others later mistook for Mansarowar. Hindu geography places the source of the Ganges at the "Cow's Mouth," an ice cave located above Gangotri, so it is possible that Andrade visited it, since he was in the neighborhood. But almost 200 years would pass before Westerners fully surveyed the upper reaches of the Ganges, discovering its source to be among the numerous glaciers and snow-fed streams near Badrinath and Kedarnath. Six years after Andrade's journey, another Jesuit priest, Francisco de Azevedo, followed his route to the Mana La, which he crossed, and then continued north through the harsh plateau country all the way to Leh, the capital of Ladakh. Azevedo returned a different way, via the Kulu Valley, after first crossing four high passes, including the 5,334-meter Tagalaung La, the first European to traverse that high and remote terrain.

In the same year that Andrade was exploring the western Himalaya, two other Jesuit priests, Fathers Cacella and Cabral, were exploring the eastern sectors of the range. They crossed the Assam highlands into the Raidak Valley of Bhutan in March 1627 and then struck northward through the Chumbi Valley into Tibet. Cabral reached the Tibetan town

of Shigatse in 1628 and returned to Calcutta via Nepal along an unknown route. The mountains of Nepal were more fully explored a few decades later by Fathers Grueber and Orville, who journeyed from Lhasa to Kathmandu in 1661. Their route took them through the Kham highlands of Tibet and down the Bhote Kosi River in Nepal, putting them within sight of Mount Everest. Jesuit explorations of the Himalaya slowed for a time,

curtailed by Vatican decree, and more than half a century passed before another Jesuit priest, Father Ippolito Desideri, undertook extensive travels in the Himalaya. Desideri's route led him through the Pir Panjal Range of present-day Himachal Pradesh and into Kashmir, then across the Zoji La to Ladakh. Desideri continued eastward through Tibet, reaching Lhasa two years later. After spending five years in the Tibetan capital,

The Ganges River issues full force from an ice cave in the snout of the Gangotri Glacier at Gaumukh— the physical source of the Ganges and a place of great spiritual significance.

Exploration and Travel 159

## » EARLY EXPLORATION BY EUROPEANS

The nineteenth century was a period of intensive European exploration throughout the Himalaya. Much of this reconnaissance was prompted by colonial interests—to secure the boundaries of empires and to survey the resources of the mountains. The British were especially keen to mark the northern limits of their Indian colony in advance of border movements by China and Russia.

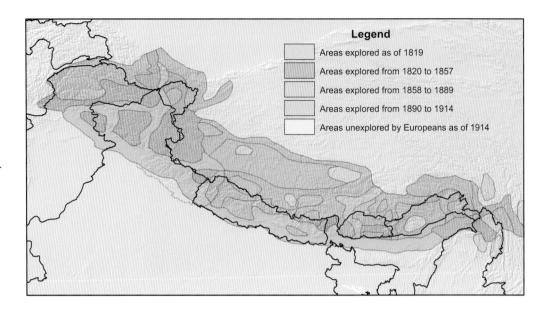

EARLY EXPLORATIONS: REGIONAL SETTING

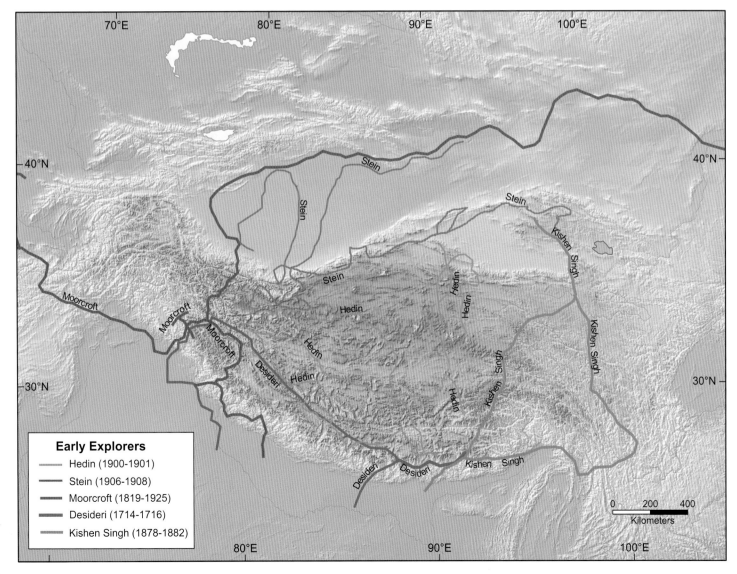

**Early Explorers**
- Hedin (1900-1901)
- Stein (1906-1908)
- Moorcroft (1819-1925)
- Desideri (1714-1716)
- Kishen Singh (1878-1882)

Desideri returned to Delhi via the central Himalaya, passing through Kathmandu en route.

Despite the Jesuits' amazing travels, pieced together by members of their order, they provided remarkably little geographic information about the Himalaya. They were determined men obsessed with their missions, but they lacked formal training in surveying or topography. By most accounts, the early Jesuit travelers also lacked the knack for narration, restricting their diary entries to the barest bones of description. Desideri provided what is probably the most extensive account of his travels, focusing mainly on Tibet, but even his descriptions lacked reliable geographic references. The extensive travels of Fathers Grueber and Orville through the central Himalaya are recounted in a few letters they sent to their Jesuit superiors, but the priests omit any reference to the majestic terrain. Apart from a few measurements of latitude made along the way and some scattered statistics about local populations, Grueber's letters give little account of either Himalayan topography or ethnography. The dearth of information from the Jesuits was so acute, in fact, that the first European maps of the Himalaya, published in 1735 by the geographer D'Anville for his atlas of China, relied instead on the maps and surveys of the Chinese lamas.

So it is that, despite the lengthy sojourns of the Jesuit priests during the seventeenth and eighteenth centuries, as well as the travels of the many traders and adventurous characters who preceded them, most commentators date the beginning of scientific exploration in the Himalaya to the sponsored expeditions of British India. These geographic adventures had a decidedly imperial significance. In 1803 Charles Crawford, officer in charge of the British residency to Kathmandu, returned to India from Nepal with a large-scale map of the Kathmandu Valley based on his direct observations and a small-scale map of the entire country based on secondhand information he had collected from travelers. In 1804 Crawford completed a new set of surveys of eastern Nepal that included the calculated heights of many of the great snowy peaks. In 1807 two British army officers, Captain William Webb and Captain Felix Raper, undertook to discover the source of the Ganges River. Their perambulations provided a set of detailed maps at a 1-inch scale, as well as journals filled with keen observations on culture, topography, and the estimated heights of many of the observed summits. In all, Crawford's cartographic contributions to the British Empire were substantial.

A few early explorations from elsewhere in the Himalaya were documented, including those of the Jesuits. Some regional historians point to the detailed descriptions of Kashmir provided by Frenchman François Bernier, who visited the valley in 1663, as the beginning of the modern era of exploration. Others suggest that the maps produced in 1808 by the British dispatch to Kabul, Sir Mountstuart Elphinstone, constitute the best starting point for modern Himalayan exploration. But it is the surveys of Craw-

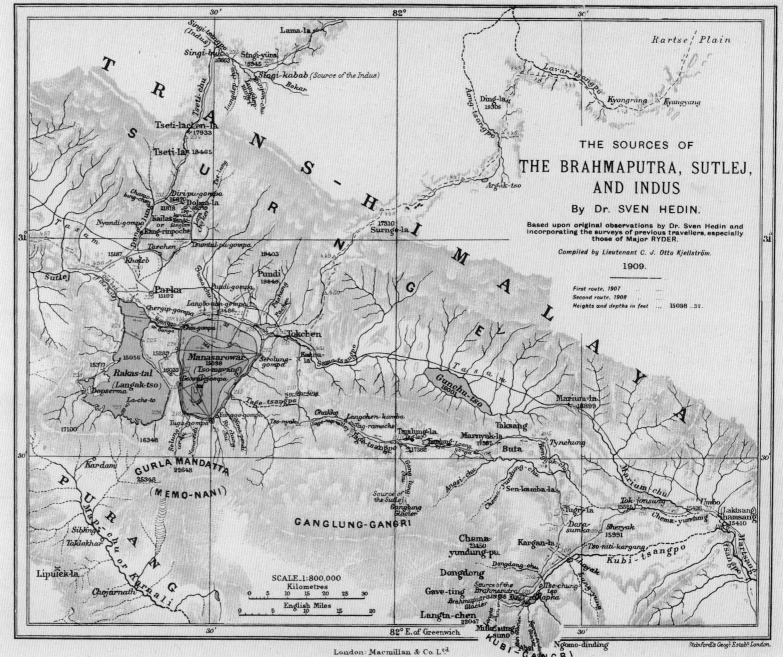

SVEN HEDIN'S MAP OF MAJOR HIMALAYAN RIVER SOURCES

One of the most famous Himalaya explorers was Sven Hedin, who crisscrossed the western Himalaya and journeyed into the harsh landscapes of the Takla Makhan desert in the north. Hedin also was one of the first Europeans to explore inside Tibet. His early map defines the Himalaya by the bends of the Indus and Brahmaputra rivers but provides scanty cartographic detail about the interior of the mountains.

Exploration and Travel 161

EARLY EXPLORERS' ROUTES IN EAST-CENTRAL HIMALAYA

Much of the exploration of the Himalaya during the eighteenth and nineteenth centuries followed established trade routes in the mountains. As a result, the European explorers brought little new information to local inhabitants, but they made the mountains known to the Western world.

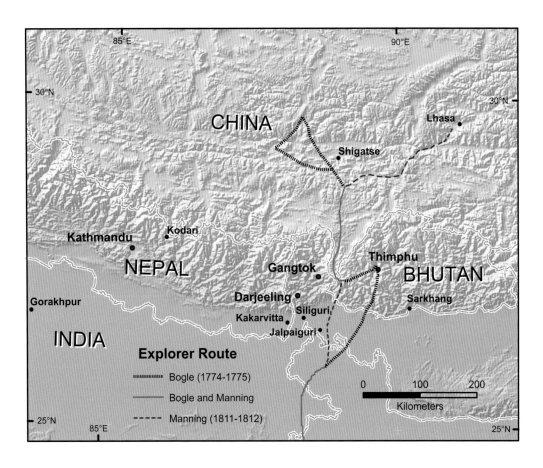

## The British Explorers

The first official expedition by the British into the Himalaya had a mercantile purpose, with the intention of opening avenues of trade between India and Tibet. In 1774 the British East India Company sent one of its revenue collectors, George Bogle, to Bhutan and thence to Tibet via the Chumbi Valley. His mission was to make contact with local rulers and to establish a viable travel route from Calcutta to Lhasa. Bogle spent quite a lot of time in Bhutan, noting its unique culture and the natural qualities of the land before venturing on to Tibet, where he struck a lasting friendship with the Panchen Lama in the town of Desheripgay. Bogle never achieved a treaty with Tibet, and he never met with the Tibetan ruler he had set out to negotiate with, so in those respects, his journey was a failure. However, Bogle returned to India with a great deal of information about Bhutan and Tibet and with a fairly good idea about prospective trade routes across the trans-Himalaya. Little attention was given to his proposals until thirty-six years later, when his accounts provided the travel route particulars for the eccentric adventurer Thomas Manning. Manning had no official standing in the East India Company, which considered his erratic behavior of little benefit to the empire, and his journey to Lhasa was strictly personal. He, too, provided little geographic information of practical interest to the British. Shortly after Manning's return to India in 1812, both Bhutan and Tibet closed their borders to all foreigners, sealing their geographic mysteries for over a century, and the British turned their attention to exploring the western frontiers of the Himalaya.

In 1819 the East India Company sent William Moorcroft, manager of the company's horse farm in Bihar, on a trading expedition to Central Asia, purportedly to buy breeding stock. A year later, his party crossed the Sutlej River on inflated buffalo skins, followed the Beas River north to Zanzkar, and pioneered a difficult route across the Bara Lacha pass to Leh in Ladakh, crossing territory previously unknown to Europeans. Moorcroft spent three years in Ladakh awaiting permission from the Chinese to proceed northward into Central Asia. He used

ford in Nepal, and the expeditions of Webb and Raper in the Kumaon Range, that go on record as being the most notable. Although they were not the first men to explore the mountains, they were the first to do so with an official imperial sponsor, securing them a prominent place in the annals of Himalayan exploration. The absence of reliable cartographic information, at a time when the imperial boundaries of Britain, China, and Russia were converging on the Himalaya, sparked a period of intensive exploration and mapping in the mountains during the early nineteenth century. Much of this effort came from the British, who viewed the Himalaya as the northern frontier of their Indian empire, and whose sponsored explorations were intended to demarcate the empire's boundary and to discover the wealth contained in the lands within the imperial reach.

his time to advantage by exploring much of the region, including the high, remote valleys around Leh and the barren deserts of the Changthang. Finally, still lacking the necessary permits from China or permission from his company superiors, Moorcroft left Ladakh and proceeded west across the Zoji La to Kashmir, where he spent the next ten months. From there, he continued on to Afghanistan, where he died mysteriously on August 27, 1825, at the age of sixty, ending what is probably the greatest journey ever undertaken in the Himalaya.

Moorcroft was already an old man when he began his Himalayan expedition, and one can only imagine the stamina required to cross the high snowy passes, ford rivers, and

Old settlement quarters of Leh, the capital of Ladakh and an important trading center in the western Himalaya.

> » EARLY EXPLORERS' ROUTES IN WESTERN HIMALAYA

William Moorcroft, a British trader and spy, was one of the most intrepid explorers of the nineteenth century. His journeys into the western Himalaya spanned almost a decade, and he spent several years living in Ladakh. He provided the British government with a wealth of geographic knowledge about the hostile regions of the upper Indus River.

Villagers at a temple ceremony in Ladakh exhibit customs and costumes that have changed little over several centuries.

traverse some of the most difficult and hostile terrain in the world. His entourage included surveyors and cartographers, suggesting that his travels were more than a mere trading expedition. In fact, Moorcroft was a spy for the British government. The account of his travels, contained in his diary and published later as the book *Travels in the Himalayan Provinces*, provides lengthy but rather prosaic details about village customs, mountain agriculture, and topography. It reads much like a gazetteer, or a compendium of facts and observations, rather than the daring tale of adventure it must have been. Moorcroft soon became somewhat of a legend among travelers in the western mountains, partly because of the unknown circumstances of his death, which some attribute to murder by thieves and others to poison by double-crossing authorities. Moorcroft's own speculations about the British Empire's use of the western Himalayan territory, which he proposed openly in letters to his superiors, were largely ignored, and his fame as an explorer dimmed in the wake of those who followed him into the mountains.

The colonial government of India remained intent on mapping the mountains and their geographic qualities to bolster its northern defenses against the advances of the Russians and the Chinese. Early explorations of the Himalaya must be viewed in light of this cartographic interest. Adventurers such as Moorcroft were employed as secret agents by the British in what has come to be called the "Great Game," a period of es-

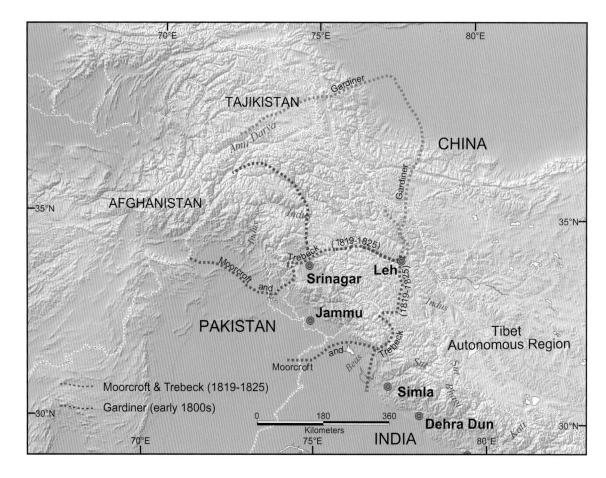

pionage and subterfuge centering around the shifting boundaries of South Asia's political frontier. The men recruited to explore the mountains were often eccentric and idiosyncratic loners, self-indulgent and vain, but they possessed enormous persistence and an unusual stamina for solitude and hardship. Their journeys were conducted in disguise, mainly to outwit the local warlords, who forbade foreigners in their unwelcoming lands. Traders, artists, and scientists with keen skills of observation made up a good portion of the exploration parties. Other members were simply adventurers who relished spending years at a time in the harsh outposts of Asia and who, by virtue of their language and observational abilities, were in a position to gather important geographic knowledge. Reading the accounts of their travels, it is not hard to imagine that the political ruse of their assignments simply added to the intrigue of travel in the High Himalaya.

John Keay's book *When Men and Mountains Meet* provides a stirring account of the exploits of European explorers in the western part of the Himalaya. He recounts the journeys of Jacquemont, who spent four months in Kashmir in 1831 collecting plants, rocks, and stuffed animals; of Joseph Wolff, who traveled through Kashmir and Ladakh as an insane religious fakir; and of Godfrey Thomas Vigne, who drew detailed maps of the western mountains in 1835 and hid them in the canvases of his remarkable paintings of the Kashmir Valley. The early European explorations centered mainly on the western ranges, including the Karakoram and Hindu Kush. Much of the central Himalaya was off-limits; the rulers of Nepal and Bhutan kept their countries closed to outsiders. In the eastern sectors of the range, the ferocity of the mountain tribes, the lack of trails, and the soggy weather kept the British at bay.

It was imperative, however, that the British learn as much as they could about the entire range of mountains. Under the direction of surveyor and cartographer William Lambton, they had embarked in 1800 on the Great Trigonometrical Survey (GTS), a mathematical gridiron of triangulated points that would permit the calculation of all locations and heights in India. The Himalaya remained a great white swath of ignorance in the emerging map of South Asia. The survey was expanded in the 1820s and 1830s under the direction of Sir George Everest, who as surveyor-general of India took the GTS into the High Himalaya. It was a remarkable feat of surveying. According to geographer Kenneth Mason, whose book *Abode of Snow* provides a classic account of the Himalaya explorations, Everest's efforts made it possible "to fix with considerable accuracy the positions and heights of the highest summits without visiting them. No man before or since has done so much for the geography of Asia."

Given the dangers of travel in the Himalaya—especially for foreigners, who had to contend with hostile local authorities and could easily be picked out of a crowd of natives—the British began to employ Indians to survey the mountains. The native explorers, known as pandits, were trained in simple cartographic arts. They were taught how to calculate distances by using a measured pace, measure angles with a compass and sextant, and determine the gradient of hill slopes with a clinometer. They used prayer wheels, filled with 100 beads instead of the customary 108, as mechanical counters to keep track of their measurements, and they learned how to determine elevation by noting the boiling point of water. The pandits often traveled as pilgrims, sometimes along the ancient religious routes, and endured tremendous hardship along the way. The information they obtained was indispensable to the colonial mission, but they received few accolades among European society, and with minor exceptions, their explorations have been largely overlooked.

The Indian explorers gained access to many parts of the Himalaya that were closed to Europeans. In 1865 Nain Singh

Sir George Everest, superintendent of the Great Trigonometrical Survey of India, 1823–1843; surveyor-general of India, 1830–1843. (Courtesy of the Royal Geographical Society)

Wandering holy man, known as a *sadhu*, follows a pilgrim trail near the Bhagirathi Peaks.

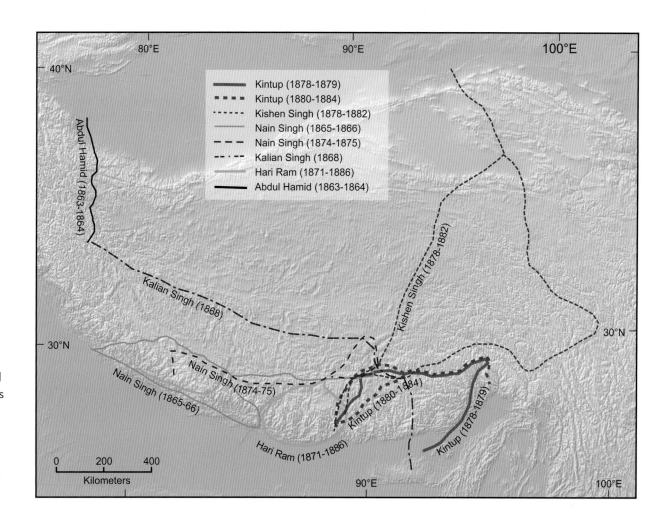

» ROUTES OF INDIAN AND TIBETAN EXPLORERS

Native guides, interpreters, and porters regularly accompanied the early British explorers. A few of the Hindu guides, known as pandits, became renowned Himalayan explorers in their own right. Their journeys, though often overlooked by the Western media, provided important geographic information for many later explorations, including those conducted by Ian Baker and colleagues in the mid-1990s that led to the discovery of a large waterfall along a remote stretch of the Tsangpo Gorge.

traveled west across Kumaon and Garhwal into Nepal as far as Kathmandu, whence he proceeded north to Tibet. Ten years later he traveled from Lhasa to Gauhati, India, crossing Bhutan from north to south. In the late 1870s Kishen Singh explored much of Sikkim and the mountains north of present-day Arunachal Pradesh, in the eastern sector of the Himalaya. During 1871–1872 Hari Ram completed an arduous route through Nepal and Sikkim that encircled Mount Everest, which must have looked like a tiny peak on the horizon when it was not obscured by the surrounding summits.

One of the most celebrated native explorers was a tailor from Sikkim named Kintup, known in the official British archives as KP. In 1878 he first served as an assistant to the pandit Nem Singh on an expedition in eastern Tibet. Their goal was to follow the Tsangpo River from the Tibetan Plateau all the way down to the Brahmaputra Valley in Assam. The pair never made it through the difficult Tsangpo Gorge, leaving unexplored a 150-kilometer stretch of the river. In 1880 Kintup embarked on further explorations of eastern Tibet, this time accompanying a Chinese lama under the employ of Henry Harman, a British captain in charge of surveying the upper Assam region. The lama proved treacherous, though, and gave Kintup to local officials in the Kham village of Tongjuk Dzong in exchange for a horse. Kintup escaped the authorities after

Exploration and Travel  167

The explorer Kintup. (Courtesy of the Royal Geographical Society)

seven months of slavery and spent the next four years traveling through the remote upper Tsangpo region. His report, finally published in 1889, disclosed a forbidden landscape of lush forests, deep gorges, and secretive waterfalls, a place known to Tibetans as the mythical Pemako. Kintup's descriptions of this remarkable region, known previously to only a handful of pilgrims, would eventually inspire the explorations of later Europeans, including Tibetologist Frederick Bailey in 1911–1912 and botanist Frank Kingdon Ward in 1923. Most recently, Kintup's journey was the backdrop for expeditions led by Ian Baker and Kenneth Storm Jr., who in the mid-1990s undertook a series of journeys that were sponsored in part by the National Geographic Society, leading to the eventual reconnaissance of the Hidden Falls of the Tsangpo Gorge.

The geographic information provided by the colonial explorers and the pandits was invaluable in filling in many of the blank spots on the British maps of South Asia. Their land surveys helped situate the Himalaya within Sir Everest's great mathematical spheroid of India. In 1834 surveyors under his direction were in Dehra Dun, lining up the mountains in the western range. They gradually worked east and west from that point, determining the elevations of the great peaks by measuring their angles from known heights at calculated distances. In 1852 the Bengali chief surveyor working under the direction of Sir Andrew Waugh measured an obscure mountain in eastern Nepal named Peak XV at 8,840 meters (actual height, 8,850 meters). The story goes that he marched into the survey office and breathlessly exclaimed, "Sir, I have discovered the highest mountain in the world." Although many people have refuted that scenario, it captures a dramatic moment in an otherwise long and tedious cartographic effort. In 1865 the peak was named Mount Everest, in recognition of the work of the former surveyor-general of

Indo-Russian survey station, Pamir, 1913. (Courtesy of Maria Brothers Bookstore, Simla).

India. In 1858, K2, the second-highest mountain on earth, was surveyed and calculated to be 8,621 meters (actual height, 8,611 meters). One by one, the high summits were triangulated and measured with amazing accuracy. It is impossible to overestimate the challenges encountered by the survey crews working in the Himalaya. For wages as low as a dollar a month, they climbed some of the highest mountains on earth armed with theodolites and plane tables, setting their instruments atop the peaks, measuring the major summits, and filling the intervening distances with information gained from their trail explorations. The surveyors encountered malaria, vipers, and hostile tribes in the lower elevations and had to contend with treach-

Bhagirathi Peak I (6,856 meters) in the Garhwal region.

Climbing party en route to Mount Everest relies on yaks to carry heavy loads into the high elevations. The expedition here is passing through Namche Bazaar, a major Sherpa settlement in the Khumbu region that serves as headquarters for Sagarmatha National Park.

erous glaciers, avalanches, and menacing blizzards amid the high peaks. In the end, they succeeded in demarcating some of the most magnificent mountains on earth.

## The Mountain Climbers

The first climbers in the Himalaya were members of the British survey teams. By 1862, the indefatigable workers of the Great Trigonometrical Survey of India had reached the summits of more than thirty-five peaks exceeding 6,000 meters elevation. Those climbs were no easy feats, given the lack of technical mountaineering experience and the fact that the team members carried heavy survey equipment up every summit. On one of the climbing expeditions in the Zanzkar Range in 1860, a native Indian surveyor attained an altitude of 7,025 meters, carrying a signal pole to the top of Mount Shilla. He held the world altitude record for several decades, but we do not even know his name. One of the great mountaineers of all time was Captain Godwin-Austin, who led the mountain surveys in northern Pakistan during the early 1860s. Godwin-Austin mapped many of the glaciers of the Karakoram Range and surveyed the topography of the high peaks around K2, climbing many of the lesser peaks in the process. The highest triangulation stations of the survey, however, were set in Kashmir in 1862 by W. H. Johnson, all of them situated at elevations above 6,000 meters.

For the survey teams, climbing the Himalayan peaks was all in a day's work, and their achievements were not recorded as

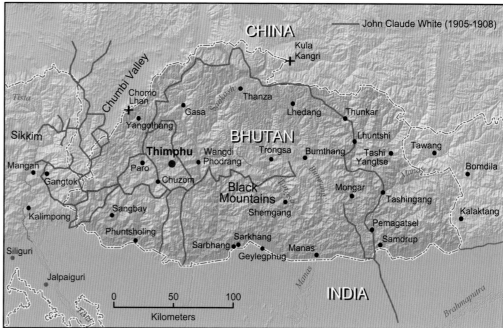

mountaineering exploits. In 1885 the British sent Francis Younghusband on a mission to investigate the area north of Gilgit in Hunza, the first official expedition with the explicit objective of mountain exploration. His efforts provided a wealth of geographic information not only about the mountains and glaciers but also about the passes and travel routes across the northern Indian borderlands. Although he is thought of as the first Himalayan mountaineer, Younghusband was a colonial officer in the British survey, and his alpine exploits had a decidedly imperial mission.

The first person to arrive in the Himalaya for the explicit and sole purpose of climbing mountains for sport rather than for the advancement of scientific or political knowledge was W. W. Graham, who attempted in 1883 to climb first Kanchendzonga in Sikkim and then Nanda Devi in the Kumaon Range. His assaults on the Himalayan giants were unsuccessful, but Graham spent much of the year traipsing up the various lesser peaks in the region, often attaining heights over 6,000 meters. His casual accounts are filled with many errors of fact and observation, however, making it difficult to determine their validity. Nonetheless, Graham's adventures were creditable enough to gain the attention of the Royal Geographical Society in London, which, based on the strength of his Himalayan field studies, eventually sponsored many of the mountain climbers who succeeded him.

The patronage of the Royal Geographical Society grew out of the perception that mountaineering had much to offer geographic exploration and provided the appealing combination of adventure and science. The first Himalayan expedition sponsored by the society was led by alpinist Martin Conway, whose 1892 adventures in the Karakoram sought, among other things, to discover "the limit to which qualified mountaineers can climb without being stopped by the rarity of air." Conway's climbing party included four Gurkha soldiers on loan from a British regiment. The strength and acumen of these Nepalese mountaineers were first proved on Conway's Karakoram

« Expeditionary porter in the western Himalaya.

⌃ BHUTAN EXPLORATION: JOURNEYS OF JOHN CLAUDE WHITE

The interior of Bhutan was closed for much of history, jealously guarded by its dharma kings from outside influences. This mysterious land was known as the Dragon Kingdom, and it was not until the twentieth century that Westerners, led by John Claude White, began to explore its hidden reaches. Bhutan still retains its aura of mystique and inaccessibility.

Exploration and Travel   171

(these pages) FIRST ASCENT ROUTES OF 8,000-METER PEAKS

All the 8,000-meter peaks in the Himalaya have been climbed successfully. The initial ascents mark important watersheds in the history of mountaineering, not only capturing the adventure of the climb but also ushering in new routes, alpine techniques, equipment, and climbing strategies. Many of the ascent routes forged by the first successful climbing expeditions still guide contemporary mountaineers, whose numbers in the Himalaya grow each year, bringing money to cash-strapped regions and crowding the treacherous upper slopes.

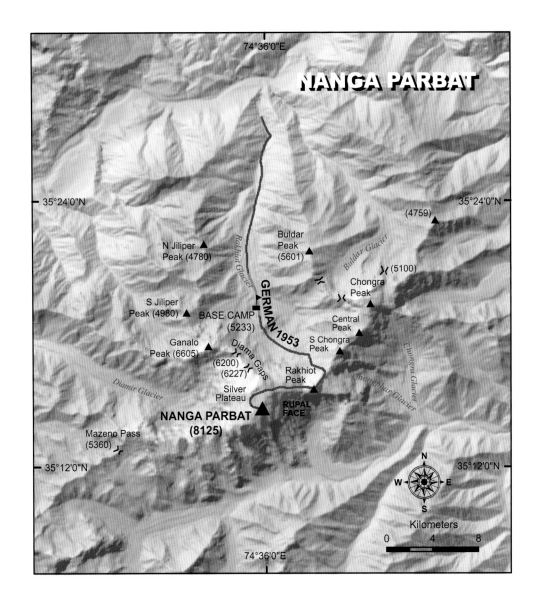

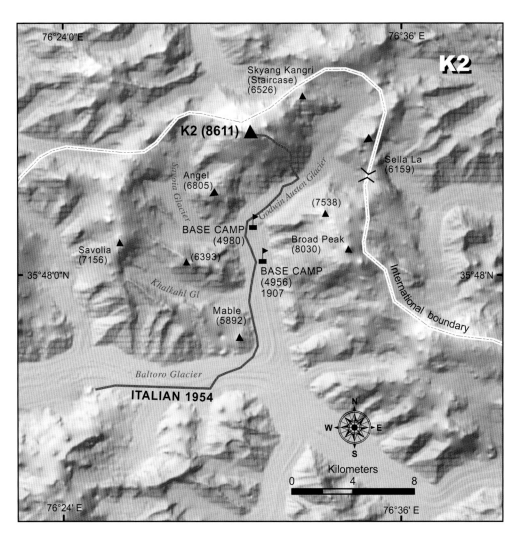

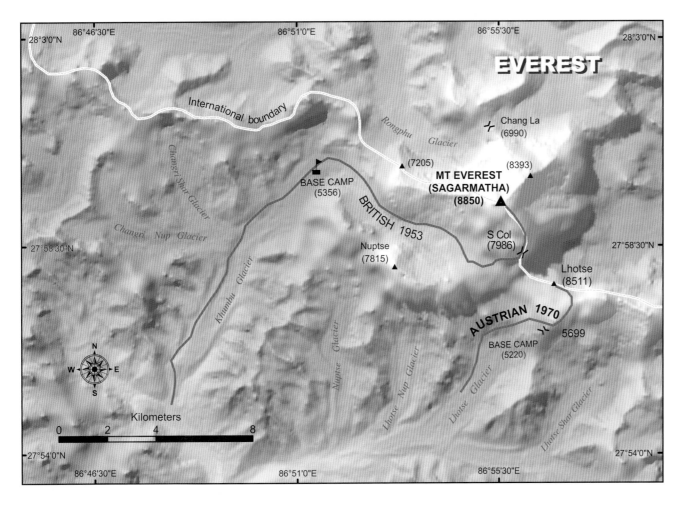
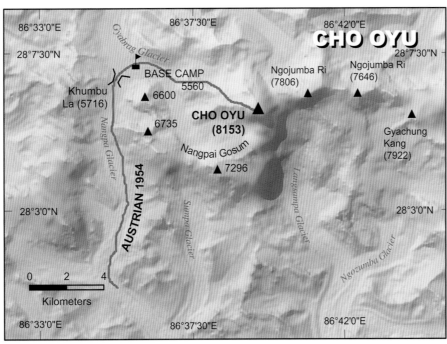
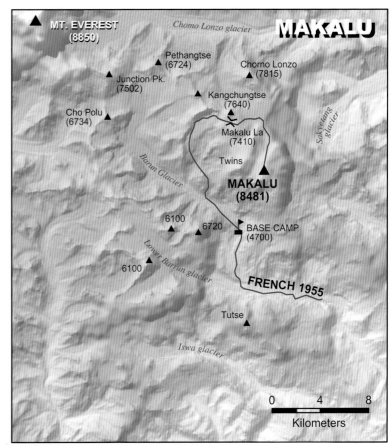

Lamayuru Monastery in the Zanzkar Range.

expedition, leading to their future role in alpine exploration throughout the High Himalaya. The reputation of the Nepalese Sherpa guides would culminate some sixty years later, at precisely 11:30 AM on May 29, 1953, when Tenzing Norgay and New Zealand climber Edmund Hillary stood together at the summit of Mount Everest.

In 1895 a team of British alpinists arrived in India with the intention of climbing Nanga Parbat. The expedition ended in tragedy when some of its members, including the accomplished climber A. F. Mummery, were killed in an avalanche, but it stands as the first truly expeditionary attempt to climb one of the giant peaks of the Himalaya. The organization of

Mummery's Nanga Parbat expedition, and its backing by the Royal Geographical Society, gave it the logistical support and credibility that would mark the later successful Himalayan summit assaults. The turn of the twentieth century found a record number of mountaineers attacking the high peaks. T. G. Longstaff attempted an eastern approach to Nanda Devi (7,816 meters) in the Kumaon Range in 1905, reaching 6,400 meters elevation before turning back due to the condition of the snow. In 1909 the Italian duke of the Abruzzi, Luigi Amedo, organized an assault on K2 supported by 300 local porters. His attempt was short-lived due to the danger of the terrain, but the Italian mountaineer turned his attention to nearby Bride Peak and attained a record height of 7,503 meters. In 1912 C. F. Meade bivouacked at 7,138 meters up Mount Kamet above the Pindar River, the highest overnight encampment then on record.

The early climbers focused mainly on the mountains of the western range, especially those in Kashmir and Kumaon. Nepal and Bhutan remained closed, their high peaks off-limits

Monks in a ritual procession at Tabo Monastery in the Spiti Valley.

⌃ Herders' trail in the remote Zanzkar mountains serves as a trekking route for adventurous tourists.

« Mountaineer Appa Sherpa hails from the famed Khumbu region of Nepal and holds the Guinness World Record for Mount Everest ascents—fifteen successful summits.

to foreign mountaineers. In the east, an international group of mountaineers arrived in Sikkim in 1905 to climb Kanchendzonga. That attempt ended in disaster after the group gained an altitude of 6,705 meters and an avalanche buried several members of the climbing party. In attempting the summits of K2, Nanga Parbat, Nanda Devi, and Kanchendzonga, as well as countless lesser peaks, the early European climbers learned a great deal about the topography of the High Himalaya, as well as about the physiological conditions encountered by climbers above 6,000 meters elevation. The growing body of information about acclimatization would prove necessary for later expeditions. The early expeditions also initiated a siege-style approach to mountaineering, by which a series of encampments at progressively higher altitudes, provisioned by porters, laid the foundation for a final summit assault from the highest camp. This method was adopted by almost all the later successful climbing expeditions.

With a growing body of knowledge about the mountains, and the increasing expertise and technical skills of the mountaineers, it was only natural that the grand Himalayan prize, Mount Everest, would come into play. The main problem was that most of the mountain lay in the forbidden country of Nepal. In 1921, however, the Royal Geographical Society formed an Everest Committee to look into the prospects of climbing Everest from the northern Tibetan side. Warming relations between Britain and the Lhasa government provided optimism that permission to do so would be forthcoming. After permission was granted, a reconnaissance of the mountain was made in 1921, and the first assault of the summit took place in 1922. The 1922 expedition did not reach the top of Everest, but climbers E. F. Norton, George Mallory, and Howard Somervell attained an altitude of 8,225 meters, the first persons to break the 8,000-meter barrier. It was as much a psychological accomplishment as a physical one. Two years later, in 1924, the Tibetan government sanctioned another attempt on the mountain. It too was unsuccessful and cost the lives of two of the world's great climbers, George Mallory and Andrew Irvine. It remains somewhat of a mystery whether Mallory and Irvine actually reached the top of Everest; their ice ax was discovered at about 8,500 meters, and the two were reportedly observed at 8,600 meters, but most people assume that they did not reach the pinnacle of the world.

Despite several climbers reaching heights above 8,000 meters, only two of the famous Himalayan giants, Kamet (7,761 meters) and Nanda Devi (7,822 meters), were conquered prior to 1950. Kamet's summit was reached in 1931 by a climbing party led by Frank Smythe and including Eric Shipton. H. W. Tilman and N. E. Odell climbed Nanda Devi in 1936. With the opening of Nepal to mountaineers in 1949, most of the world's 8,000-meter peaks came into the climbers' gaze. The first of these to be summitted was Annapurna (8,091 meters) in 1950 by members of a French expedition led by Maurice Herzog. He described being on top of the mountain in his famous book *Annapurna:* "A fierce and savage wind tore at us. We were on top of Annapurna! 8,075 meters, 26,493 feet. . . . The summit was a corniced crest of ice, and the precipices on the far side which plunged vertically down beneath us, were terrifying, unfathomable. . . . Above us there was nothing." Herzog's measurements were a bit off—he was actually standing on higher ground—but the expedition stands as one of the most inspired feats of mountaineering in the chronicles of the sport. Three years later, in 1953, Mount Everest (8,850 meters) and Nanga Parbat (8,125 meters) were climbed. Thereafter, the 8,000-meter giants fell in succession until 1964, when the Chinese finally scaled the last one, Shisha Pangma (8,046 meters).

An entire library of mountaineering literature exists that describes the famous and not-so-famous assaults on the Himalayan peaks, with the climbers themselves providing some of the most stirring narratives. De Filippi's *Karakoram and the Western Himalaya,* an erudite, cerebral account of mountaineering in northern Pakistan, and Herzog's impassioned book *Annapurna,* chronicling his team's climb of the mountain, are two very different kinds of books, but both stand as early classics in the field. Many of the books written by climbers record their determination in overcoming all manner of physical and mental challenges in order to reach the tops of earth's highest places. They portray the bravery and comradeship of the climbing team members, including the porters and Sherpa guides, as well as the fractious nature of mountaineering expeditions made up of idiosyncratic and headstrong people. A good compilation of Himalayan climbing history is Ian Cameron's *Mountains of the Gods,* and Louis Baume provides a complete chronology of the successful climbs to the summits of the 8,000-meter peaks in his book *Sivalaya.*

After the early successes, in which the summits of the Himalayan giants were conquered, the mountaineers proceeded toward new frontiers of climbing: scaling massive rock walls, establishing new routes, completing solo climbs, and reaching the summits without the use of bottled oxygen. The legion of famous Himalayan climbers grew remarkably during the 1970s and 1980s, led by people such as Chris Bonington, the first to

Trekkers' lodge in Khumjung village, Nepal.

Exploration and Travel 177

(this spread) **HIMALAYA TREKKING AND MOUNTAINEERING REGIONS**

Ancient travel and trade routes in the Himalaya now serve as the basis for modern trekking excursions. Popular destinations for trekkers include the Mount Everest and Annapurna regions in Nepal and the Zanzkar Range and Indus Valley in Ladakh.

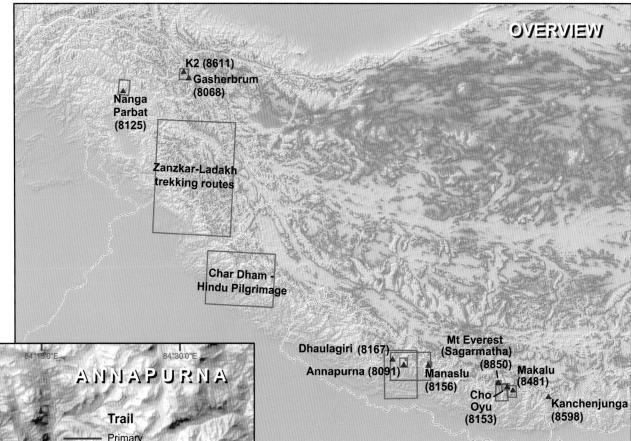

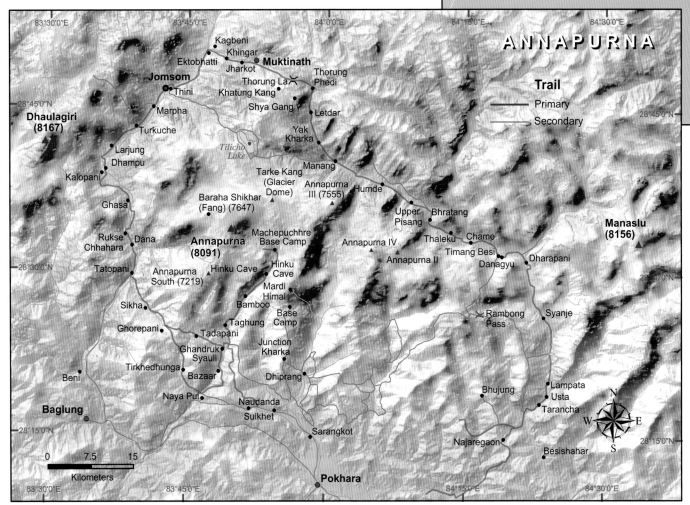

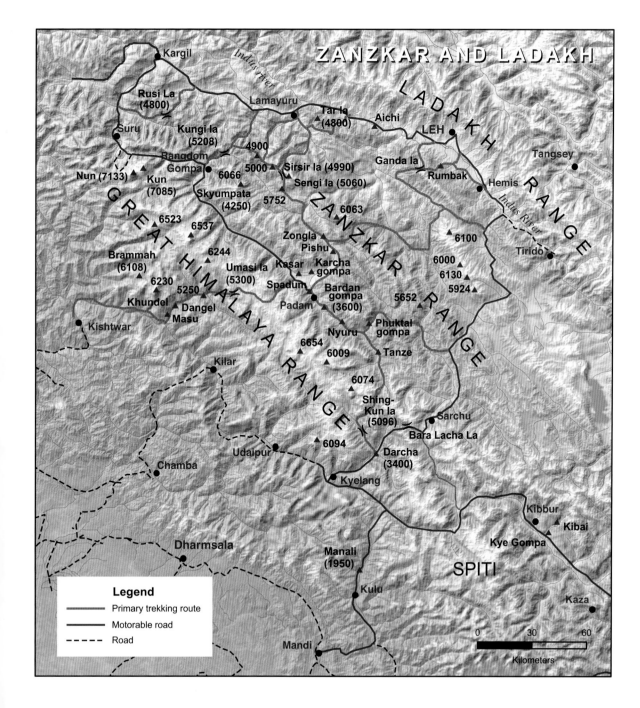
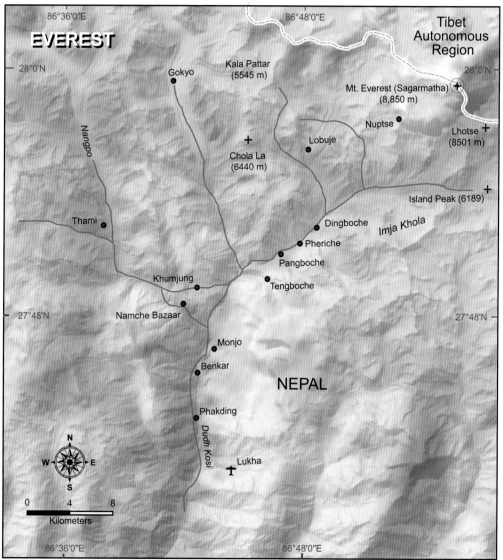

introduce giant rock wall climbing in the Himalaya, and Reinhold Messner, who climbed the major Himalayan summits alone and without supplemental oxygen. The modern breed of Himalayan mountaineer has expanded most recently to include people with relatively little mountaineering experience. Guided climbs for amateurs are now commonplace in the Himalaya, including on Mount Everest, provoking considerable controversy about whether they are appropriate. The clients pay great sums of money to reach the highest places in the world, but they often arrive without proper training or motivation. A tragedy occurred in 1996 when several client-climbers, as well as their accomplished guides, lost their lives in a storm atop Everest. Acrimony as well as heroism prevailed throughout that tragedy. Some fear that guided climbs will eventually diminish the challenge of the mountains; others contend that the tremendous difficulty of reaching the high places will always prevail and that mountaineering in the Himalaya will remain one of the world's most arduous and impassioned sports.

In addition to relating the dizzying dangers of the heights, the climbing classics describe the beauty of the mountains and the pleasures encountered on the approach treks to reach the mountains' flanks. These descriptions often gush with idyllic imagery of the glorious countryside and quaint native villages, providing a siren song for the legions of people who began visiting the Himalaya in the late 1960s for the purpose of walking along the trails described by the mountaineers. Though the summits may remain out of bounds for all but the most accomplished alpinists, less demanding treks among the lower altitudes are manageable and immensely enjoyable for many people. During the 1970s and early 1980s, many places in the Himalaya were developed for trekking-based tourism.

## Trekkers and Modern Tourism

Adventure trekking as a form of travel and exploration in the Himalaya actually dates to the mid-1800s, when British families on holiday in the Indian hill stations of Kashmir, Simla, and Darjeeling first began spending time in the mountains walking, fishing, hunting, and picnicking. Trekking in the modern sense, however, is linked to the indefatigable efforts of Jimmy Roberts in Nepal in 1966, when he established the country's first trekking company and began leading guided tours into the Nepalese countryside. Since then, the appeal of trekking among Western tourists has skyrocketed, with upward of a million people a year now visiting the Himalaya. Adventure trekking is best established in Nepal, where it constitutes the mainstay of the country's travel and tourism industry. There are countless tourist agencies in Nepal, and they, along with the many international travel outfits operating there, have funneled hundreds of thousands of trekkers annually into the kingdom in recent years. Most of the international trekking occurs along well-developed trail circuits, often contained within the territories of national parks or conservation areas. The two oldest trekking circuits are the Solukhumbu trails near Mount Everest and the Annapurna highlands north of Pokhara Valley. Other designated trekking destinations in Nepal include the Langtang Valley north of Kathmandu, Dolpo in the western trans-Himalayan region, and relatively new circuits in the Mustang, Makalu, and Kanchendzonga regions. These regions offer treks ranging in duration from a few days to several weeks.

The Mount Everest area is included in Sagarmatha National Park, which was established in 1976 to protect 1,148 square kilometers of territory. The park's centerpiece, of course, is the world's highest mountain, but it also contains numerous peaks over 7,000 meters, expansive forests of rhododendron and hemlock, endangered wildlife, and a human population of 3,500 Sherpa villagers. Most trekkers in Sagarmatha National Park follow a route that begins at the airstrip in Lukla and proceeds north for a two-day walk along the Dudh Kosi River to the market village of Namche Bazaar, which acts as a gateway to higher elevations. Beyond Namche, trekkers can proceed past the famous Thyangboche Buddhist monastery en route to the Everest Base Camp or take an alternative route to Gokyo Lake, with its outstanding views of Everest and the surrounding

Trekkers on an adventure tour in the Spiti Valley camp below the ancient fortress remains of Dankar Monastery.

peaks. Either way, the Everest trekking trails command spectacular views of many of the world's highest mountains and pass among numerous Sherpa villages, whose strong cultural traditions persist amid the influx of foreigners. The economic importance of trekking in the Mount Everest region is registered by the fact that 75 percent of the Sherpa population is employed in tourism-related jobs.

In recent years, the greatest number of tourists—over 50,000 a year, on average—visit the Annapurna highlands, which was fully opened to trekkers in 1977. The famous Annapurna circuit takes about twenty days to complete and introduces trekkers to a wide range of landscapes and cultures. It completely encircles the Annapurna massif and includes a rigorous climb up the 5,416-meter Thorung La pass on the north side of Annapurna, bringing trekkers within reach of alpine country that is normally visited only by accomplished mountaineers. The full range of environments encountered in the Annapurna circuit—from subtropical forests to snowy peaks—the diversity of native plants and animals, and the sprawling countryside filled with mountain villages make this a notable route

Exploration and Travel 181

Suryakund Falls in Uttaranchal is a popular destination for trekkers, pilgrims, and tourists.

in the canons of Himalayan trekking. The entire Annapurna circuit is contained with the boundaries of the Annapurna Conservation Area, Nepal's largest protected area (7,629 square kilometers), which is managed by the King Mahendra Trust for Nature as a multiple-use park.

Many trekkers are members of expeditionary groups, which are properly outfitted with guides, porters, cooks, and equipment. However, it is also possible to travel as individuals, in pairs, or in small groups, relying on local villages for accommodations and food. This so-called tea-house trekking, unique to Nepal, is possible because the villages are conveniently spaced along the trails and maintain their ancient tradition of inn keeping. The small-scale nature of the tea-house approach to trekking allows travelers to interact with villagers and arrange unique itineraries. In many ways, it closely resembles the serendipitous forms of pilgrimage and exploration that have historically defined travel in the Himalaya.

Trekking has facilitated the modern exploration of Nepal, especially in the remote areas of the kingdom that have only recently been opened to foreigners. These destinations are located mainly within the Great Himalaya and usually adjoin Tibet along Nepal's rugged northern border. Included in this list of places are the Kanchendzonga Conservation Area, situated along Nepal's border with Sikkim; Makalu Barun National Park, which adjoins the Mount Everest region; the upper reaches of Mustang in the north-central part of the kingdom; and the Manaslu Eco-Tourism Area. These trekking spots represent some of the most isolated places in the Nepalese Himalaya and generally require a certain amount of self-sufficient, expeditionary-style travel; they provide a unique sense of discovery and are popular among more experienced trekkers. The logistical challenge of travel is heightened in such regions by their inaccessibility; they lack the amenities found along the more popular trekking circuits, and trekkers are required to obtain travel permits and other documentation in order to visit them.

Although Nepal is most famous worldwide for its Himalayan trekking, the western parts of the range in India also contain a large number of trekking regions. These became more popular in the early years of the twenty-first century when civil unrest in Nepal escalated, and security concerns led trekkers to seek other Himalayan destinations. Kashmir Valley has long been a favorite destination, with many treks leading into the surrounding mountains. The British hiked a great deal in Kashmir during the early 1900s, when it served as a summer resort area for colonial families. Kashmir was popular among foreign tourists in the 1970s and early 1980s, but the escalation of civil violence during the 1990s and the tensions between Pakistan and India along the mountain border have diminished the prospects for tourism in the western part of the state. In the eastern part of Kashmir, however, the Buddhist regions of Zanzkar and Ladakh are relatively peaceful and offer outstanding trekking possibilities. Unlike in Nepal, trekking in Zanzkar and Ladakh requires considerable advance planning. The trails are rough, river crossings are often required where no bridges exist, villages are scarce and ill equipped to meet the needs of travelers, and many stretches of the trail require trekking parties to be self-sufficient in terms of food for several days. For these reasons, most visitors hire horsemen and guides, whose services are invaluable in navigating the high, unmarked terrain.

Quite a few trekking circuits exist in Zanzkar, most of which traverse wide, isolated valleys and cross high passes over 5,000 meters, following age-old trading routes. The landscape is dominated by open vistas where the Suru Valley divides the Great Himalaya and the Zanzkar Range; elsewhere one can find savage gorges, snowy peaks, stone villages, and Buddhist *gompa*s. A newly constructed road connects the town of Kargil, which is on the main road between Srinagar and Leh, with the village of Padum, which forms the nexus of the region. As a result, the central part of Zanzkar is changing rapidly. Elsewhere, though, the district is accessible only by trail, and its inhabitants still cling to ancient ways of life. Trekkers in Zanzkar normally follow one of three routes. A twelve-day circuit from Padum north to Lamayuru proceeds via five high passes to a crossing of the rugged Zanzkar Range. A ten-day trek leads south from Padum into the Great Himalaya, crosses the 5,100-meter Shingola La pass, and descends to the village of Darcha, located on the main road from Manali to Leh. A third route goes from Padum to Leh via a seven-day stretch across uninhabited highlands, climbing three 5,000-meter passes along the way. All three trekking routes include high altitudes, difficult ascents, and challenging river crossings. The passes are snowbound for much of the year, and the rivers become swollen and treacherous from snowmelt during the early summer months.

Located to the northeast of Zanzkar is the district of Ladakh, known as the "land of high passes." Ladakh first opened to tourism in 1974 as one of the remotest regions in the Indian Himalaya. Most visitors begin their explorations in Leh, the ancient trading and monastic center situated along the Indus River amid towering, snowy peaks. The British explorer Moorcroft lived in Leh during the 1820s, and many of his descriptions of this ancient place—physically dominated by the hulking old palace that hangs over the town—still ring true, despite advancing modernity and a heavy Indian military presence. One of the numerous treks that begin in or near Leh is the popular one from nearby Spitok through the Markha Valley. The destination of this one-week trek—Hemis Monastery—is one of the most important Buddhist sites in Ladakh. It is the principal monastery of the Drukpa order, and the head monk of Hemis oversees all the monasteries in the entire Ladakh and Zanzkar region.

Ladakh's strategic location on the main trading routes between Central and South Asia has been a decisive factor in its political and economic life. Nowadays, the modern adventure traveler adds a new twist to the ancient mix of camel caravans, bazaars, and pilgrims that ply Ladakh's age-old travel routes. It is not much of a stretch for modern trekkers to imagine themselves as early explorers, discovering long-lost temples set deep within a Shangri-La landscape, despite the burgeoning and sophisticated tourism industry. Both Zanzkar and Ladakh hold great promise for personal exploration by foreigners due to their isolated locations, ancient customs, and spectacular terrain.

The Indian states of Himachal Pradesh and Uttaranchal, located on the southern side of the Great Himalaya, also offer a number of fine trekking regions. The celebrated Kulu Valley, the "valley of the Gods," is a center of numerous trekking and mountain exploration circuits, ranging from short excursions among highland pastures to week-long expeditions into the Great Himalayan National Park, whose headquarters is located in the valley. The most popular treks in Kulu generally take three to five days and often lead to glacial lakes, high shepherd pastures, or prominent summits located amid the high ranges surrounding the upper valley. Popular destinations include the alpine lake Beas Kund, the glaciers below the summit of Deo Tibba, and Chatru village, which is reached after crossing 4,330-meter Hampta Pass. Although relatively short treks by Himalayan standards, the routes in the Kulu Valley are challenging, with steep ascents through forests and alpine pastures that require time for acclimatization.

The upper Sutlej River in Himachal Pradesh provides access to several noteworthy trekking regions that were only recently opened to foreigners. The Baspa Valley, renowned for its beautiful landscape and picturesque villages, is reached by traveling northeast from the hill station of Shimla along the Hindustan-Tibet road as far as Karcham village and then proceeding south along the Baspa River. The Kinnaur Kailash Range looms above the Baspa Valley near the Tibet border. An ancient pilgrimage route circumambulates Bharawan Peak in the Kailash Range and provides a trekking circuit from Thangi village over the Lalanti Pass to Chitkul village and then north again to the Sutlej Valley. Beyond the juncture of the Sutlej and Spiti rivers are the isolated upper valleys of Kinnaur and Spiti. The trekking circuits in the Spiti Valley cross over the Great Himalaya into the Pin Valley, famous for its rugged terrain, endangered ibex herds, and snow leopards. The trails into the Pin Valley are difficult, with dangerous river crossings and glacial passes over 5,600 meters. For hardy and experienced trek-

>> Tiger's Nest Monastery, Bhutan.

kers, though, they provide access to some of the most remote and challenging terrain in the entire western Himalaya.

Several trails proceed south out of the upper Baspa Valley, crossing the Bandrapunch Range in the Great Himalaya via the Lamkhaga (5,284 meters), Khamilogo (5,151 meters), or Borsu (5,360 meters) passes and into the Yamuna River watershed. These great treks sweep down onto the holy mountains around the ancient pilgrimage sites of Yamunotri and Gangotri in the Garhwal region of Uttaranchal. Many of the trekking routes in Garhwal lie in the monsoon realm, making them hard to manage during the summer months due to heavy rains and swollen rivers. The most popular treks in Garhwal follow ancient pilgrimage trails to the mouths of sacred rivers, to meditation caves, and to other sacred Hindu places. The destinations of these treks include Gaumukh and the Gangotri Glacier above Gangotri Temple, the Kedarnath temple complexes, and alternative sources of the Ganges River at Madmaheshwar and the Khatling Glacier. Two additional trekking circuits of note in Garhwal are the upper reaches of the Bhyuntar River, known as the "valley of the flowers," which was first visited and described by Frank Smythe in the 1930s during his explorations of the terrain south of Mount Kamet, and the Nanda Devi Sanctuary. Both these places are designated national parks, and access for trekking is restricted for reasons of environmental conservation.

In the eastern Indian Himalaya, both Darjeeling and Sikkim offer possibilities for trekking and exploratory travel. The Darjeeling treks tend to be relatively short excursions out of town, past the tea estates planted by the British in the 1930s, and through the lush eastern forests. The center for trekking in Sikkim is the Yuksam-Dzongri Plateau south of Mount Kanchendzonga, which borders Nepal. This is a restricted area, and trekkers must be accompanied by licensed professional guides. Bhutan lies to the east of Sikkim and also offers excellent but restricted opportunities for trekking. Bhutan allows only a small number of visitors each year, and most of them do not travel into the remote countryside, so for those that do, the sense of discovery and exploration is great. The state of Arunachal Pradesh, located in the far eastern Himalaya, provides few opportunities for trekking because most of its mountainous terrain remains off-limits to foreigners. On the Tibetan side of the eastern Himalaya, recent explorations have led to organized trekking excursions into the once-forbidden landscapes of Pemako, including access to pilgrimage sites tucked deep within the Tsangpo Gorge. Overall, mainly due to political considerations, the eastern Himalaya is much less heavily visited by foreigners than the central and western sectors and, for that reason, retains a certain mystique among avid Himalayan travelers.

Although most modern Himalayan trekking is a long way from the experiences of the early explorers, it still provides outstanding opportunities for personal discovery. It does not matter that a place has been visited before; it is always a singular experience for first-time visitors. And although many places have been thoroughly mapped, the remote spots remain largely outside the sphere of major modern influences, so that the trekking circuits provide windows into ancient lifestyles as well as spectacular mountain scenery. With its recent explosive growth, the trekking and tourism industry is bringing significant changes to the Himalaya. In some places, environmental degradation occurs as a result of litter, trail erosion, and the cutting of trees for heating and cooking fuel. The presence of Westerners in remote villages may slowly erode cultural traditions that persist in the mountains. Such impacts, of course, describe the long history of mountain exploration, of which trekking is a modern extension, and travel in general brings global influences to the high and wild places of the Himalaya.

# Sources of Illustrations

## PART ONE  The Regional Setting

General reference maps (Himalaya: regional setting, western sector, central sector, eastern sector): compiled by authors from various sources, including government maps, atlases, tourist maps, global data sets, U.S. National Aeronautics and Space Administration (NASA) satellite data.

Administrative districts in the Himalaya: International Centre for Integrated Mountain Development (ICIMOD) (boundary status as of September 2002)

Administrative regions of Indian Himalaya: adapted from Indian Himalaya: A Demographic Database 2002.

Kathmandu Valley: ICIMOD, Mountain Environment and Natural Resources Information System (MENRIS) data.

Kathmandu city: ICIMOD, MENRIS data.

Srinagar city: Dubey and Sinclair 1992.

Kashmir Valley: compiled by authors.

Pokhara central valley: ICIMOD, MENRIS data.

Pokhara town: ICIMOD, MENRIS data.

Bhutan valley settlements (Paro Valley, Thimphu Valley, Bumthang): Karan 1967; Pommaret 1991.

Gangtok city: Sikkim Department of Tourism.

Shimla town: Dubey and Sinclair 1992; Sud 1992.

Dehra Dun city: Indian Himalaya tourism map.

Biratnagar city: ICIMOD, MENRIS data.

Nepalganj city: ICIMOD, MENRIS data.

## PART TWO  The Natural Environment

Drifting continents and formation of the Himalaya: compiled by Julsun Pacheco from various sources, including Getis et al. 2000.

Geology of Zanzkar and Indus Valley: assembled from Crook and Osmaston 1994; Gaur 1993.

Himalaya satellite imagery: compiled by Julsun Pacheco with NASA data.

Geologic cross section: adapted from Hagen 1980.

Northward drift of India: assembled from various sources, including Gansser 1964; MacFarlane et al. 1999; Gaur 1994.

Generalized geology of the Himalaya: simplified from MacFarlane et al. 1999; Gansser 1964.

Geology of Kashmir: assembled from various sources, including Gansser 1964; Malinconico and Lillie 1989.

Geology of Kumaon: based on materials provided by the Wadia Institute of Himalayan Geology, Dehra Dun.

Geology of Nepal: simplified from Amatya and Jnawali 1994.

Geoecology of western mountains: adapted from Zurick 1988.

Geology of Bhutan: simplified from geological map of Bhutan, United Nations Economic and Social Commission for Asia and the Pacific in cooperation with the Department of Geology and Mines, Thimphu, Bhutan.

Himalaya mean temperature and climate: ICIMOD, MENRIS data.

Precipitation in Ladakh and the western Himalaya: adapted from Pangtey and Joshi 1987.

Nepal temperatures and humidity: ICIMOD data.

Nepal temperature and precipitation: adapted from UNEP 2001.

Orographic precipitation: compiled by authors.

Seasonal climate system: adapted from Wallen 1992.

Precipitation patterns in Nepal: ICIMOD data.

Himalaya annual precipitation: ICIMOD, MENRIS data.

Bhutan annual precipitation. adapted from *Bhutan: Seventh five year plan* 1992–1997.

Seismic hazards of the Himalaya: adapted from Gaur 1993.

Seismic hazards in Nepal: adapted from UN Disaster Management Team 2001.

Epicenters of major earthquakes: adapted from Gaur 1993.

Major earthquakes of the Himalaya: adapted from Gaur 1993.

Soil erosivity potential: adapted from Lauterburg 1993.

Kathmandu liquefaction hazard: adapted from Engineering and Environmental Geology Map of the Kathmandu Valley. Kathmandu: Department of Mines and Geology (with data from HMG Department of Hydrology and Meteorology, HMG Ministry of Housing and Physical Planning, HMG Department of Forestry).

PART THREE **Society**

Himalaya cultural regions: adapted from Ives and Messerli 1989.

Major Himalayan trade routes: compiled by authors; adapted from Zurick and Karan 1999.

Himalaya population density: adapted from Zurick and Karan 1999; based on data from Census of India, HMG Nepal Central Bureau of Statistics, Government of Bhutan Census Report; updated with information from various government censuses for 2000.

Himalaya population growth: adapted from Zurick and Karan 1999; based on data from Census of India, India District Gazetteers, HMG-Nepal, Central Bureau of Statistics (timeline for some data is less than the total period).

Himalaya ethnic groups: compiled by authors from various sources.

Ethnic groups in Nepal: compiled by authors from various sources.

Languages in Nepal: adapted from Gurung 1998.

Religions in Nepal (Hindu, Muslim, Buddhist): compiled from Nepal government statistics and adapted from Gurung 1998.

Tribal population of Indian Himalaya: adapted from Indian Himalaya: A Demographic Database 2002.

Urban population of Indian Himalaya: adapted from Indian Himalaya: A Demographic Database 2002.

Towns in Indian Himalaya: compiled and adapted from Sharma 2001.

Rural population of Indian Himalaya: Indian Himalaya: A Demographic Database 2002.

Urbanization in Nepal: adapted from UNEP 2001.

Himalayan roads: compiled by David Zurick; adapted from Zurick and Karan 1999.

Nepal road network: ICIMOD, MENRIS data.

Nepal airports: adapted from HMG-Nepal, Topographic Survey Branch map data, Kathmandu.

Communications in Bhutan: *Bhutan: Seventh five year plan 1992–1997*.

Nepal's landless and marginal farm households: ICIMOD, MENRIS data.

Tourist arrivals in Nepal: UNEP 2001.

Literacy rate: adapted from Indian Himalaya: A Demographic Database 2002.

Educational facilities in Bhutan: *Bhutan: Seventh five year plan 1992–1997*.

Access to potable water in Nepal: UNEP 2001; ICIMOD, MENRIS data.

Kathmandu Valley water quality: *Bhutan: Seventh five year plan 1992–1997*.

Health facilities in Bhutan: *Bhutan: Seventh five year plan 1992–1997*.

PART FOUR **Resources and Conservation**

Himalaya landscape regions: compiled by authors; adapted from Zurick and Karan 1999.

Himalaya land use: ICIMOD, MENRIS data.

Himalaya soil classification: ICIMOD, MENRIS data.

Himalaya cropland distribution: ICIMOD, MENRIS data.

Land use in Indian Himalaya: adapted from Indian Himalaya: A Demographic Database 2002.

Land use in Nepal: UNEP 2001; State of the Environment—Nepal 2001.

Agricultural crops in Indian Himalaya: adapted from Indian Himalaya: A Demographic Database 2002.

Nepal change in population and cultivated area: adapted from UNEP 2001.

Himalaya population and farmland: adapted from Zurick and Karan 1999; based on data from various government sources.

Himalaya annual farmland change: adapted from Zurick and Karan 1999; based on data from Government of India Ministry of Environment and Forests and Ministry of Agriculture, India District Gazetteers, HMG-Nepal, Central Bureau of Statistics, Land Resources and Mapping Project—Nepal, Government of Bhutan Planning Commission (data timeline for some districts is less than the total period).

Nepal cultivated area: ICIMOD, MENRIS data.

Nepal irrigated area: ICIMOD, MENRIS data.

Nepal sloping terraced area: ICIMOD, MENRIS data.

Chemical fertilizer use in Nepal: ICIMOD, MENRIS data.

Nepal grassland: ICIMOD, MENRIS data.

Nepal livestock and grazing area: UNEP 2001.

Himalaya forest cover: adapted from Zurick and Karan 1999; compiled from various sources, including Kawosa 1998; ICIMOD, MENRIS data.

Nepal per capita forest area: ICIMOD, MENRIS data.

Forest change in the tarai districts, Nepal: ICIMOD, MENRIS data.

Himalaya population and forest area: adapted from Karan and Zurick 1999; compiled from various government census and land-use reports for Bhutan, India, and Nepal.

Himalaya forest cover change: adapted from Zurick and Karan 1999; based on data from Government of India Ministry of Environment and Forests and Ministry of Agriculture, Land Resources Mapping Project—Nepal, Government of Bhutan Planning Commission.

Nepal population and fuelwood consumption: UNEP 2001.

Fuelwood deficit areas in Bhutan: FAO 1991.

Export of forest products, Nepal: UNEP 2001.

Nepal mineral deposits: adapted from HMG-Nepal, Department of Mines and Geology, Kathmandu.

Bhutan mineral deposits: adapted from *Bhutan: Seventh five year plan* 1992–1997.

Himalayan rivers: compiled by authors; adapted from Zurick and Karan 1999.

Stream flow in three rivers: adapted from Bruijnzeel and Bremmer 1989.

Changing course of the Kosi River, Nepal: adapted from Carson 1985.

Hydroelectric projects in the Bhagirathi River basin: compiled by authors from various Indian media sources, and from Paranjye 1988.

Hydropower development in Nepal: HMG-Nepal, Department of Topographic Survey data.

Household access to water and electricity in Indian Himalaya: Indian Himalaya: A Demographic Database 2002.

Glaciers, glacial lakes, and watersheds of Nepal and Bhutan: ICIMOD, MENRIS data.

Royal Chitwan National Park: adapted from HMG-Nepal, Department of National Parks and Wildlife Conservation, 2000. Royal Chitwan National Park and Bufferzone Resource Profile. Kathmandu.

Himalaya national parklands: compiled by David Zurick; adapted from Zurick and Karan 1999.

Protected areas in Nepal: ICIMOD data; various reports from HMG-Nepal, Department of National Parks and Wildlife Conservation.

Tarai arc conservation landscape: World Wildlife Fund-Nepal Program.

Himalaya protected area and biodiversity: compiled from Shengji 1995.

PART FIVE  Exploration and Travel

Ancient Hindu holy places: compiled from maps in Bhardwaj 1973.

Map of Kathmandu, circa 1879–1884: Gurung 1983.

Char Dham pilgrimage: compiled by authors from field study.

Pilgrimage route to Muktinath: compiled by authors from field study.

Exploration of Himalaya by Chinese travelers: compiled by authors from various sources, including Royal Geographical Society 1991; Olschak et al. 1987.

J. B. d'Anville map: Library of Congress; used by permission.

Exploration of Himalaya by Europeans: compiled by authors from various sources, including Keay 1993; Wessels 1998; Cameron 1984; Schwartzberg 1978.

Early explorers: compiled by authors from various sources, including archival records held by the India Office Library, London.

Sven Hedin map: sheet map, courtesy of Maria Brothers Bookstore, Shimla.

Early explorers' routes in east-central Himalaya: compiled from Cameron 1984.

Early explorer's routes in western Himalaya: compiled from Keay 1993.

Routes of Indian and Tibetan explorers: compiled from Waller 1990.

Journeys of John Claude White: compiled from maps provided to authors by Chencho Tsering, Wind Horse Adventures, Bhutan.

First ascent routes of 8,000-meter peaks: compiled from various sources, including maps provided by the Nepal Mountaineering Association, Kathmandu; Baume 1979.

Annapurna trekking: compiled from various sources, including "Around Annapurna" trekking map (NEPA MAPS, Kathmandu).

Everest trekking: compiled from various sources, including mapping data provided by Map Point, Kathmandu (courtesy Rajendra Shrestha).

Zanzkar-Ladakh trekking: compiled by authors from field study.

Himalaya mountaineering and trekking regions: compiled by authors.

# Bibliography

Amatya, K. M., and B. M. Jnawali. 1994. *Geological map of Nepal.* Kathmandu: KAAAS Consultancy and His Majesty's Government, Survey Department.

Baume, Louis. 1979. *Sivalaya: Explorations of the 8000-meter peaks of the Himalaya.* Seattle: The Mountaineers.

Bhardwaj, Surinder. 1973. *Hindu places of pilgrimage in India.* Berkeley: University of California Press.

Bhati, J. P. 1990. *Development strategies in Himachal Pradesh.* Kathmandu: International Centre for Integrated Mountain Development (ICIMOD).

*Bhutan: Seventh five year plan.* 1992–1997. Vol. 1. *Main plan document.* Thimphu: Planning Commission, Royal Government of Bhutan.

*Bhutan national human development report 2000: Gross national happiness and human development—searching for common ground.* Thimphu: Planning Commission Secretariat, Royal Government of Bhutan.

Bista, Dor Bahadur. 2000. *People of Nepal.* Kathmandu: Ratna Pustak Bhandar.

Bruijnzeel, L. A., and C. N. Bremmer. 1989. *Highland-lowland interactions in the Ganges Brahmaputra river basin.* Kathmandu: ICIMOD Occasional Paper No. 11.

Cameron, Ian. 1984. *Mountains of the gods.* New Delhi: Time Books International.

Carson, Brian. 1985. *Erosion and sedimentation processes in the Nepalese Himalaya.* Kathmandu: ICIMOD Occasional Paper No. 1.

Coburn, Broughton. 1997. *Everest: Mountain without mercy.* Washington, D.C.: National Geographic Society.

Cox, Kenneth, ed. 2001. *Frank Kingdon Ward's riddle of the Tsangpo gorges.* (Original text by Frank Kingdon Ward; additional material by Kenneth Cox, Kenneth Storm Jr., and Ian Baker). Suffolk, U.K.: Antique Collectors' Club.

Crook, John, and Henry Osmaston, eds. 1994. *Himalayan Buddhist villages.* New Delhi: Motilal Banarsidass Publishers.

Directorate of Economics and Statistics. N.d. *Important statistics of Himachal Pradesh.* Shimla.

Dubey, Manjulika, and Toby Sinclair, eds. 1992. *Insight guides—Western Himalaya.* Singapore: APA Publications.

Filippi, Filippo de. 1912. *Karakoram and western Himalaya, 1909: An account of the expedition of H.R.H. Prince Luigi Amedeo of Savoy, Duke of the Abruzzi.* London: Constable.

Finlay, Hugh, et al. 1997. *Nepal.* Hawthorne, Australia: Lonely Planet Publications.

Food and Agriculture Organization (FAO). 1991. *Wood energy sector analysis, Bhutan.* Bangkok: UN FAO.

Gansser, Augusto. 1964. *Geology of the Himalayas.* London and New York: John Wiley Interscience Publishers.

Gaur, V., ed. 1993. *Earthquake hazard and large dams in the Himalaya.* New Delhi: Indian National Trust for Art and Cultural Heritage.

*Geological map of the western Himalaya.* 1992. New Delhi: Government of India.

Getis, A., et al. 2000. *Introduction to geography.* Boston: McGraw-Hill.

Gurung, Harka. 1983. *Maps of Nepal.* Bangkok: White Orchid Books.

———. 1998. *Nepal: Social demography and expressions.* Kathmandu: New Era Publications.

Hagen, Toni. 1980. *Nepal: Kingdom in the Himalayas.* Berne: Kummerly and Frey Publishers.

Herzog, Maurice. 1952. *Annapurna: First conquest of an 8000-meter peak.* New York: E. P. Dutton.

Huber, Toni, ed. 1999. *Sacred spaces and powerful places in Tibetan culture.* Dharamsala: Library of Tibetan Works and Archives.

*Indian Himalaya: A Demographic Database.* 2002. ENVIS Center on Himalayan Ecology, G. B. Pant Institute, Almora.

International Centre for Integrated Mountain Development (ICIMOD). 1993. *International symposium on mountain environment and development.* Kathmandu: ICIMOD.

———. 1996. *GIS database of key indicators of sustainable mountain development in Nepal.* Kathmandu: ICIMOD.

Ives, J. D., and B. Messerli. 1989. *The Himalayan dilemma.* London and New York: Routledge.

Karan, P. P. 1967. *Bhutan: A physical and cultural geography.* Lexington: University Press of Kentucky.

Karan, P. P., and H. Ishii. 1996. *Nepal: A Himalayan kingdom in transition.* Tokyo: United Nations University Press.

Kawosa, M. A. 1998. *Remote sensing of the Himalaya.* Dehra Dun: Natraj Publishers.

Keay, John. 1993. *When men and mountains meet: The explorers of the western Himalayas, 1820–75.* Karachi: Oxford University Press.

Lauterburg, Andreas. 1993. The Himalayan highland-lowland interactive system: Do land use changes in the mountains effect the plains? In *Himalayan environment: Pressure—problems—processes.* Edited by B. Messerli, T. Hofer, and A. Wymann. Berne: University of Berne, Institute of Geography.

Lewis, Todd, and T. Riccardi Jr. 1995. *The Himalaya: A syllabus of the region's history, anthropology, and religion.* Ann Arbor, Mich.: Association for Asian Studies.

MacFarlane, A., et al., eds. 1999. *Himalaya and Tibet: Mountain roots to mountain tops.* Boulder, Colo.: Geological Society of America.

MacGregor, John. 1970. *Tibet: A chronicle of exploration.* New York: Praeger.

Malinconico, Lawrence, and Robert Lillie, eds. 1989. *Tectonics of the western Himalaya.* Boulder, Colo.: Geological Society of America, Special Paper 232.

Mason, Kenneth. 1955. *Abode of snow.* New York: E. P. Dutton.

Messerli, B., T. Hofer, and S. Wymann, eds. 1993. *Himalayan environment: Pressure—problems—processes.* Berne: University of Berne, Institute of Geography.

Messerschmidt, Don. 1992. *Muktinath: Himalayan pilgrimage, a cultural and historical guide.* Kathmandu: Sahayogi Press.

Mool, Pradeep, et al. 2001a. *Inventory of glaciers, glacial lakes, and glacial lake outburst floods: Bhutan.* Kathmandu: ICIMOD.

———. 2001b. *Inventory of glaciers, glacial lakes, and glacial lake outburst floods: Nepal.* Kathmandu: ICIMOD.

*Nepal human development report 1998.* Kathmandu: Nepal South Asia Center.

O'Conner, Bill. 1997. *The trekking peaks of Nepal.* Wiltshire, U.K.: Crowood Press.

Olschak, Blanche, Augusto Gansser, and Emil Buhrer. 1987. *Himalayas.* New York and Oxford: Facts on File.

Pangtey, Y. P. S., and S. C. Joshi, eds. 1987. *Western Himalaya.* Vol. 1, *Environment.* Nainital: Gyandaya Prakashan.

Paranjpye, Vijay. 1988. *Evaluating the Tehri Dam.* New Delhi: Indian National Trust for Art and Cultural Heritage.

Pommaret, Françoise. 1991. *Introduction to Bhutan.* Geneva: Editions Olizane SA.

Price, Larry. 1981. *Mountains and man.* Berkeley: University of California Press.

*Royal Chitwan National Park: Resource profile.* 2000. Kathmandu: HMG-Nepal, Department of National Parks and Wildlife Conservation.

Royal Geographical Society. 1991. *History of world exploration.* London: Hamlyn Publishers.

*School atlas of Nepal.* N.d. Kathmandu: Survey Department, Topographic Survey Branch.

Schwartzberg, Joseph, ed. 1978. *A historical atlas of South Asia.* Chicago: University of Chicago Press.

Sharma, Pitamber, ed. 2001. *Market towns in the Hindu Kush-Himalayas.* Kathmandu: ICIMOD.

Shengji, Pei. 1995. *Banking on biodiversity: Report on the regional consultation on biodiversity assessment in the Hindu Kush-Himalayas.* Kathmandu: ICIMOD.

Shiva, Vandana, et al. 1993. *Ignoring reason, inviting disaster: Threat to Ganga-Himalaya.* Dehra Dun: Friends of Chipko and Natraj Publishers.

Shroder, John F., ed. 1993. *Himalaya to the sea: Geology, geomorphology and the quaternary.* London and New York: Routledge.

Sill, Michael, and John Kirkby. 1991. *The atlas of Nepal in the modern world.* London: Earthscan Publications.

*State of the environment—Nepal.* 2001. Kathmandu: United Nations Environment Program.

Sud, O. C. 1992. *The Simla story.* Simla: Maria Brothers.

Thakur, V. C., and B. S. Rawat. 1992. *Geological map of western Himalaya.* Dehra Dun: Wadia Institute of Himalayan Geology.

Tilman, H. W., and T. G. Longstaff. 1937. *The ascent of Nanda Devi.* New York: Macmillan.

United Nations. 1999. *Nepal: Common country assessment.* Kathmandu: United Nations System.

United Nations Disaster Management Team. 2001. Nepal: *UN disaster preparedness response plan, part I.* Kathmandu: United Nations System.

United Nations Environment Program (UNEP). 2001. *Nepal: State of the environment.* Kathmandu: UNEP.

Wallen R. N. 1992. *Introduction to physical geography.* Dubuque, Iowa: Wm. C. Brown Publishers.

Waller, Derek. 1990. *The pundits: British exploration of Tibet and Central Asia.* Lexington: University Press of Kentucky.

Weare, Gary. 1991. *Trekking in the Indian Himalaya.* Hawthorne, Australia: Lonely Planet Publications.

Wessels, Cornelius. 1998. *Early Jesuit travellers in Central Asia.* Delhi: Book Faith India.

Wood, Frances. 1996. *Did Marco Polo go to China?* Boulder, Colo.: Westview Press.

World Wildlife Fund. 2000. *WWF in Nepal: Three decades of partnership in conservation.* Kathmandu: WWF–Nepal.

Zurick, David. 1988. Resource needs and land stress in Rapti zone, Nepal. *Professional Geographer* 40(4):428–44.

———. 1990. *The Himalayas.* In *Survey of earth science.* Edited by F. N. Magill. Pasadena, Calif.: Salem Press, pp. 1073–78.

Zurick, David, and P. P. Karan. 1999. *Himalaya: Life on the edge of the world.* Baltimore: Johns Hopkins University Press.

# Subject and Name Index

Afghanistan, 156
agriculture
  commercial, 108
  and environmental resources, 101–102
  expansion of, 28
  in Ladakh, 103
  in Uttaranchal, 91
  and women, 97
agro-pastoralism, 77, 102–103
airstrip, 89
Aksai Chin, 11
Alaknanda River, 47, 158
Alexander the Great, 156
Almora, 118
Andrade, Antonio, 158
Annapurna, Mount
  ascent of, 177
  height, 18, 35
  herders, 27
  location, 15
  trekking region, 180–181
Annapurna Conservation Area, 135, 183
apple orchards, 72
architecture, 78
Arunachal Pradesh, 12, 16
  biodiversity, 133
  climate, 52
  employment in, 91
  farmland, 103
  geology, 49
  literacy rates, 93
  towns, 94
Aryans, 70
Assam Valley, 4
Avalokiteswara, 153
*axis mundi*, 3
Azevedo, Francisco de, 158

Badrinath, 47, 153, 158
Baker, Ian, 168
Baltoro Glacier, 42
Bangladesh, 121
Bara Lacha Pass, 162
Baspa Valley, 47, 184
Baume, Louis, 177
Bay of Bengal, 50
Beas River, 125
Bernier, François, 161
*beyul*, 144
Bhakra Dam, 125
Bhutan
  agriculture, 105
  area, 16
  biodiversity, 4–5, 133
  Buddhist traditions, 16
  and colonialism, 72
  contested territory of, 11
  ethnicity, 77
  forests, 113, 119
  geology, 49
  "Gross National Happiness" index, 90
  hydropower potential, 127
  literacy rates, 92
  minerals and mining, 120
  population and farmland, 108
  travel in, 162
biodiversity, 4–5, 101
  in Nepal, 133
  in eastern region, 133–134
Black Mountains, 27, 136
*bodhisattva*, 153
Bogle, George, 162
Bonington, Chris, 177
boundaries, 11–12, 15,
  mapping, 162
Brahmaputra River, 4, 27, 38, 87
  water flow, 122
bridges, 87
Broad Peak, 38
Buddhism
  architecture, 154
  and civilization, 68
  monastic traditions of, 5
  and pilgrimage, 157
  sacred geography, 145
Bukarwal, 78
Buxa, 146

Calcutta, 39
Cameron, Ian, 177
caravans, 81, 156
celestial realm, 69
*chakra*, 153
Cherrapunji, 52
*chetras*, 145
China
  contested boundaries, 11
  maps, 161
  relations with Bhutan, 11
Chipko movement, 118
Chomolungma, 17, 136, 146
Cho Oyu, Mount, 35
*chortens*, 79
Chumbi Valley, 23, 158
Churia Hills, 28
cirques, 44
civilizations, convergence of, 5–8, 71
climate
  air temperature, 55
  effects of glaciers on, 58
  and flooding, 63
  monsoon, 50
  precipitation, 52–53
  and settlement patterns, 23
  and topography, 4–5, 50
  zonation, 50
colonialism
  archival records, 107
  and exploration, 162–164
  forest records, 113
  and resources, 72
conservation
  and forest protection, 118
  parks and, 119, 134–136
  and wildlife, 134
continental drift
  evidence for, 3
  theory of, 33
cosmology, 60
Crawford, Charles, 161
culture
  diverse patterns of, 4
  and lifestyles, 8
  traditions of, 8

*dakini*, 153
dams
  and hydropower, 125
  and indigenous people, 125
  in Bhutan, 127
  in Nepal, 126
Dang Valley, 28, 46
D'Anville, Jean Baptiste, 158, 160
Darjeeling, 96, 108
  trekking in, 186
Dehra Dun, 46, 120, 168
Desideri, Ippolito, 159
*dharamasala*, 152
Dhaulagiri, Mount
  height, 18, 35
  location, 15
  herders, 27
  massif, 46
dialects, 15
Dolpo, 18, 70
*dun* valley, 28, 46
*dzong*, 17

earthquakes, 62–63
East India Company, 39, 162
economic development, 65
  and national wealth, 90
  resource frontiers, 28
  and roads, 85
  and tourism, 180
education, 92–93
elevation gradient, 5
Elphinstone, Mountstuart, 161

ethnicity, 76
Everest, George, 17, 165, 168
Everest, Mount
  ascent of, 174, 176
  conservation, 135–136
  exploration, 159
  geology, 48
  guided climbs of, 180
  height, 3, 17, 168
  mountaineering, 176–177
  national park, 180
  religious importance of, 146
  trekking near, 180–183
exploration
  and boundary demarcation, 162
  by British, 162
  by foreigners, 156
  mapping, 162
  survey and, 168–169

farming systems, 101–103
farmland, changes in area of, 107–108
fault zone, 45, 48
floods, 60, 124
  and dams, 125
  glacial lake outburst, 45
fodder, 115
forest
  area, 112–113
  conditions, 112
  conservation, 118
  decline in, 112
  degradation, 115
  in Nepal, 113
  and population density, 28
  types, 112
  in western region, 113
  and wood exports, 118
fossils, 3, 39, 44, 48
fuelwood, 115

Ganesh Himal, 121
Ganges River
　exploration, 161
　and pilgrimages, 69, 153
　source, 153, 158
　water flow, 122
Gangetic Plain, 4, 46
　floods, 64
Gangotri, 47, 153, 158, 186
Gangotri Glacier, 186
Gansser, Augusto, 39
Garhwal, 15, 47, 48, 62, 71
　commercial logging, 118
　dams and hydropower, 125
　farmland, 107
　migration, 82
　population growth, 73, 110
Gasherbrum, 38
gemstones, 42, 46
Genghis Khan, 157
geology
　and age of mountains, 42
　in central sector, 48
　continental drift, 33
　in eastern sector, 49–50
　and exposed strata, 46
　faults, 45–46
　and fossils, 39
　mapping, 39
　and mountain building, 33–35, 44
　and seismic disturbance, 33
　tectonic structures, 38, 42
　in western sector, 46
glacial lake, 61
glaciers
　and climate modification, 55
　diminishing size, 123
　and lake outbursts, 45
　location, 44
　mapping, 170
　and river flow, 27
　size, 41–42

globalization, 8
　and communications, 89
　impacts of, 138–139
　and migration, 82
　and modernity, 68
Godwin-Austin, Captain, 170
Gosainkund, 146
Gokyo Lake, 180
Gondwanaland, 33
gorges, 41, 49
Gorkha, 71
governance, 96–97
Graham, W. W., 171
Grand Pilgrimage, 152
Great Game, 71, 164
Great Trigonometric Survey, 165, 170
Gurung, 78
Gyala Peri, 49

habitat destruction, 133
Hagen, Toni, 39, 48, 120
hazards, natural, 60–65
health
　and disease, 94–96
　and health services, 95
Helambu, 145
Hemis Monastery, 184
Hemis National Park, 134
Herzog, Maurice, 177
Hidden Peak, 38
Hillary, Edmund, 174
hill station, 85, 180
Hilton, James, 144
Himachal Pradesh
　communications in, 89
　population, 15
　territory, 15
　trekking in, 184
Himalaya
　biodiversity, 133
　civilizations of, 68
　climate, 50–59

　contested boundaries of, 11
　cultures, 8
　early migrations into, 70
　environmental hazards, 60–65
　exploration, 156
　extent, 4
　forests, 112–115
　geological history, 33–35
　geological zones, 46–50
　glaciers, 123
　images, 23
　mountaineering, 170
　population, 73–76
　religious views of, 3
　river valleys, 41
　sacred geography, 145–147
　sectors, 4, 5, 11–12, 15–17
　and seismic disturbances, 63
　social conflict in, 68
　and thrust sheets, 39
　and tourism, 180
　urbanization, 81
　vertical dimensions, 5
Hinduism
　architecture, 154
　and civilization, 68
　and pilgrimage, 146, 157
Hindu Kush, 4, 156
Hispar Glacier, 42
Hsuan Tang, 156
human rights, 97, 140
Hunza Valley, 85
hydropower, 27, 61, 85, 89
　in Bhutan, 127
　and electricity generation, 125
　in Nepal, 125–126

ice age, 44
India
　and colonialism, 71
　contested boundaries of, 11
　military, 85

　migration from, 82
　and plate movements, 41, 42
　and Vedic traditions, 68
Indus River, 4, 38, 61, 115
　water flow, 122
Indus-Yarlung suture, 42, 47
International Centre for Integrated
　Mountain Development, 141
International Year of the Mountain, 141
irrigation, 27, 103
　and terraces, 28
Irvine, Andrew, 176
Islam, 5, 68

Jamnotri, 153
Jesuits, 157
*jhum*, 113–114
Jomoson, 55

K2, Mount, 35, 169
Kailas, Mount, 47, 60, 68, 145, 151
Kali, 60
Kali Gandaki Gorge, 3, 41
　air pressure and wind, 55
　hydropower potential, 126
　roads in, 86
Kamet, Mount, 175
Kanchendzonga (Kachenjunga),
　Mount, 15, 27, 35, 77, 145
　first ascent of, 171
Kanchendzonga National Park, 135
Kao-seng-tsan, 35
Karakoram
　geology, 34–35
　glaciers, 34–35
　mapping, 170
　mountaineering, 171
　peaks, 3
Karakoram Highway, 85
Kashmir
　civil unrest in, 96

　culture, 76
　economic stagnation in, 91
　geology, 46
　location, 11
　mapping, 170
　militarization, 12
　mountaineering, 175
　paintings of, 165
　travel in, 163
Kathmandu, 62, 63, 84, 160
Kathmandu Valley, 46, 49, 121
　architecture, 154
　earthquake potential, 62
　history, 70
　pollution, 84
　rivers, 124
　water table, 128
Keay, John, 165
Kedarnath, 153, 158
Khardung Pass, 85
Khumbu, 70
King Mahendra Trust, 183
Kintup, 167
Kohistan, 46
Krol Mountains, 48
Kula Kangri, 18
Kulu Valley, 85
　forests, 115
　trekking in, 185
Kumaon, 15, 48, 82
　population increase, 110
　travel in, 167
Kun Zum La, 23
Kyrgyzstan, 4

Ladakh, 4, 17, 153, 158
　agriculture, 103, 110
　culture, 79
　exploration, 162–163
　geology, 46–47
　irrigation, 110
　military roads, 19, 85
Lamayuru, 183

Subject and Name Index 197

landslides, 60–61
Langmoche Glacier, 61
Laitsawa Pass, 23
land degradation, 28
landlessness, 97
landscape
    degradation, 28
    description, 18
    and geology, 42
    model of, 5
Langtang, 41
Leh, 158, 162, 184
Lepcha, 77
Lhasa, 159, 176
livestock grazing
    and pasture expansion in Bhutan, 133
    and soil erosion, 63
Lotshampha, 96
Lumle, 55

Machapuchare, Mount, 61
Mahabharata, 145, 152
Mahabharat Lekh, 27, 45–46
Mahakali River, 15, 47
Makalu, 35
Makalu Barun National Park, 183
Mallory, George, 176
Manasarowar, 146, 158
Manaslu, 35
Manas Wildlife Sanctuary, 137
*mandala*, 3, 154
Manning, Thomas, 162
Maoist insurrection, 96, 97
maps of Himalaya, 17, 158, 161
    and cartographic surveys, 168
Marco Polo, 157
Mason, Kenneth, 165
migration, 77, 81–82
    of Nepalese into Bhutan, 110
minerals
    and mining, 120
    in Ladakh and Zanzkar, 120
monastery, 24, 180, 184

Mongols, 157
Monpa, 77
monsoon, 50
    and air pressure, 50
    and precipitation, 50–51
    and river flow, 124
Monserrate, Anthony, 158
Moorcroft, William, 162–163, 164, 184
mountain passes (*la*), 19, 23
mountaineering, 27
    in central region, 177
    history, 170
    techniques, 176
    in western region, 174
Muktinath, 146, 153
Muslims, 69
Mussoorie, 120
Mustang, 18, 70, 183
Myanmar, 50

Nainital, 48
Namche Barwa, 4, 38, 50, 123, 145, 152
Namche Bazaar, 61, 180
Nanda Devi, 152, 171, 175
Nanga Parbat, Mount, 4, 12, 33, 46–47, 60, 152
    mountaineering, 174
nappe sheet, 41, 42
National Geographic Society, 168
national parks, 118, 119
    and conservation, 133, 134–137
    trekking in, 180
    and wildlife, 131
natural hazards, 60–65
Nepal
    agriculture, 106
    climate, 50, 55, 58
    early empire, 71
    ethnicity in, 76–77
    exploration, 159
    farmland change, 108

    forest conditions, 113
    glaciers, 123
    population growth, 73
    roads, 85–86
    urbanization, 84
Newar, 78
newspapers, 89–90
Norgay, Tenzing, 174
Nubra Valley, 85

Padmasambhava, 145
Padum, 183
Pahari, 70, 76, 82
Pakistan
    contested boundaries, 11
    exploration, 156
    Indus districts, 12
    population, 76
    tectonic regions in, 38
Pamir, 4
Panchen Lama, 162
*pandits*, 165
parks
    in Bhutan, 136–137
    in Nepal, 135
    and policy, 138
    and tourism, 138
    in western region, 134–135
Pashupatinath, 153
Pasighat, 88
Pemako, 145, 168, 186
petty kingdoms, 71
pilgrimage, 146, 149, 151–153, 157
pilgrims, 149, 151–153, 154, 156
    and espionage, 165
Pin Valley National Park, 135
Pir Panjal, 27
Pokhara Valley, 49
    floods, 61
population, 73
    density, 108
    growth, 28, 73–76
    impact of migration on, 82
porters, 193

poverty
    and development, 90
    and employment, 91
prayer flags, 79, 151
prayer wheels, 165
protected areas, 134

railways, 88–89
rainshadow, 19
    effect on precipitation, 55
Rapti River, 124
refugees, 96
resettlement scheme, 107
rice terraces, 128
river systems, 27, 122–127
roads
    in Arunachal Pradesh, 87
    in Bhutan, 86
    environmental damage caused by, 87
    military, 19, 85
    and modernization, 86
    in Nepal, 86
ropeways, 88
Royal Bardia National Park, 133
Royal Chitwan National Park, 133, 135
Royal Geographical Society, 171, 175, 176

sacred geography
    and pilgrimage, 145–146
    and tribal people, 69
Sagarmatha, 17, 135, 146, 180
*saligram*, 153
salt pans, 44
scripture circuit, 153
shamanism, 78
Shambala, 145
Shangri-La, 144, 184
Sherpa, 27, 78, 97, 145, 180–181
Shiva, 60, 68, 146
Shivaling Peak, 47
Siachen Glacier, 42, 85

Sikkim
    area, 15
    empire, 71
    farmland, 103, 111
    literacy rates, 92
    passes, 23
    plateaus, 18
    population, 73
    sacred geography, 145
    trekking in, 186
    urban areas, 84
    wealth, 90
silk route, 72, 85, 157
Simla, 48, 108
Singh, Nain, 165
Siwaliks, 28, 39, 45, 49
    cropped area in, 107
slash-and-burn farming, 8
snowfields
    and river discharge, 122
soil, 101
solar radiation, 55
Spiti, 19, 47, 79, 135, 184
Spiti Valley, 24
Storm, Kenneth, Jr., 168
Subansiri River, 111
Survey of India, 120
surveyors, 164
Sutlej River, 47, 162
Swat Valley, 156

Tang Valley, 145
Takla Makan Desert, 157
tantric, 71
tarai, 28, 76, 82, 86, 107
    timber cutting in, 118
tectonic basin, 39, 46
tectonic uplift, 23, 39, 42
    and environmental hazard, 60
Tehri Dam, 125
telecommunications, 89
terrain, 5, 17
territorial rights, 69
Tethys Sea, 3, 23, 33, 39, 45

Thakali, 78
*thanka*, 154
Thimphu, 86, 90
Thorang La, 181
Thyangboche Monastery, 180
Tibet, 4
    exploration, 167
    Jesuit travel in, 160
    landscape, 18
    spiritual center, 145
Tibetan society, 70
Tibet Plateau, 17–18, 39, 42
    climate, 55
    exploration, 158, 167
    sacred geography, 153
Tien Shan, 156

Tistha River, 18, 124
toponym, 17
tourism, 138
travel
    adventure, 162
    and espionage, 164–165
    exploratory, 156–157
    by Jesuits, 157, 160
    religious, 152
trekking
    in Great Himalaya Zone, 27
    impacts of, 186
    in Nepal, 180–181
    routes, 180
    styles, 183
    in Zanzkar and Ladakh, 183

Trishuli, 41
Tsangpo Gorge, 4, 49, 167, 186
Tsangpo River, 49, 167

urbanization, 81, 84–85, 108
    and water resources, 128
Uttaranchal
    employment in, 91
    formation of, 12
    population, 15

vertical uplift, 44
Vigne, Godfrey Thomas, 165

Wadia Institute of Geology, 120
Waialeale, Mount, 52

Ward, Frank Kingdon, 168
water
    contamination, 130
    and glaciers, 123
    hydropower, 125, 127
    irrigation, 128
    as resource, 121
    and snow melt, 123–124
wildlife, 130–131
    in Bhutan, 133
    poaching, 131
    threats to, 133
women's rights, 97

Yarlung-Tsangpo River, 17, 123

Younghusband, Francis, 171
Yunnan, 153

Zanzkar, 4, 153
    agriculture, 110
    climate, 24
    extension of Tibetan empire into, 70
    geology, 47
    passes, 23
    roads, 27, 85
    survey of, 170
    trekking in, 184
Zanzkar River, 24
Ziro Valley, 111

# Map Index

| Name | Page | °N Latitude | °E Longitude |
|---|---|---|---|
| Adarsha Nagar | 26 | 28.05 | 81.61 |
| Adarsha Nayatole | 26 | 26.49 | 87.28 |
| Alchi | 179 | 34.21 | 77.20 |
| Aksai Chin | 6, 10 | 35.28 | 79.20 |
| Alak | 6 | 30.55 | 79.12 |
| Alaknanda | 6 | 30.27 | 78.95 |
| Almora | 6, 10, 13 | 29.60 | 79.65 |
| Amahi | 26 | 26.50 | 87.26 |
| Amala Bisauni | 22 | 28.26 | 83.98 |
| Amarnath Cave | 21 | 34.20 | 75.44 |
| Amarsimha Chok | 22 | 28.21 | 84.00 |
| Amintara | 22 | 28.18 | 83.99 |
| Amu Darya | 7 | 37.15 | 72.74 |
| Anantnag | 6, 10 | 33.75 | 75.19 |
| Anarwala | 25 | 30.38 | 78.07 |
| Anchar Lake | 21 | 34.15 | 74.77 |
| Annapurna | 6, 10, 13 | 28.40 | 84.42 |
| Annapurna II | 178 | 28.53 | 84.13 |
| Annapurna III | 178 | 28.61 | 84.02 |
| Annapurna IV | 178 | 28.54 | 84.08 |
| Annapurna South | 178 | 28.53 | 83.80 |
| Api | 6, 10, 13 | 30.00 | 80.93 |
| Arakot | 6, 10 | 30.99 | 77.83 |
| Arun | 7, 13 | 27.30 | 87.20 |
| Arunachal Pradesh | 7, 14 | 28.06 | 94.07 |
| Aryanagar | 26 | 26.42 | 87.29 |
| Ashokchok | 26 | 26.44 | 87.29 |
| Asonli | 7, 14 | 28.72 | 95.56 |
| Assam Valley | 7, 14 | 27.71 | 94.93 |
| Ashley Hall | 25 | 30.38 | 78.07 |
| Awantipura | 21 | 33.89 | 75.00 |
| | | | |
| B. P. Chok | 22, 26 | 28.22 | 83.99 |
| Babugaun | 26 | 28.04 | 81.62 |
| Badaharbot | 22 | 28.26 | 83.99 |
| Badrinath | 6, 10, 13 | 30.83 | 79.45 |
| Bagale Tol | 22 | 28.22 | 83.99 |
| Bagar | 22 | 28.24 | 84.00 |
| Baglung | 6, 10, 13 | 28.27 | 83.57 |
| Bagmati River | 20 | 53.76 | 89.13 |
| Bahrabise | 7, 13 | 27.79 | 85.90 |
| Baidam | 22 | 28.21 | 83.96 |

| Name | Page | °N Latitude | °E Longitude |
|---|---|---|---|
| Bairabana | 26 | 26.41 | 87.29 |
| Bajhapata | 22 | 28.21 | 84.02 |
| Bajrabarahi | 20 | 53.65 | 89.08 |
| Bajrajogini | 20 | 53.78 | 89.31 |
| Bakhari | 26 | 26.43 | 87.26 |
| Balaju | 20 | 53.78 | 89.05 |
| Balambu | 20 | 53.75 | 88.96 |
| Ballupur | 25 | 30.38 | 78.07 |
| Baltistan | 6, 10 | 35.54 | 75.28 |
| Balwahi | 26 | 26.44 | 87.30 |
| Bamboo | 178 | 28.46 | 83.86 |
| Bandipura | 21 | 34.40 | 74.64 |
| Baneswor | 20 | 53.74 | 89.11 |
| Bangemuda | 20 | 53.76 | 89.06 |
| Bankegaun | 26 | 28.05 | 81.62 |
| Baraha Shikhar | 178 | 28.59 | 83.79 |
| Bara Lacha La | 6, 10, 179 | 32.83 | 77.36 |
| Baramula | 6, 10, 21 | 34.19 | 74.34 |
| Bardan gompa | 179 | 33.35 | 76.96 |
| Bargachhi | 26 | 26.46 | 87.28 |
| Barhaghare | 22 | 28.18 | 84.01 |
| Basanta Baithak | 22 | 28.21 | 83.97 |
| Bastatol | 26 | 26.48 | 87.27 |
| Battachok | 26 | 26.47 | 87.29 |
| Batule Chaur | 22 | 28.26 | 83.98 |
| Beas | 6 | 32.07 | 75.58 |
| Belaspur | 26 | 28.07 | 81.62 |
| Beni | 178 | 28.34 | 83.56 |
| Benkar | 179 | 27.75 | 86.71 |
| Besishahar | 178 | 28.22 | 84.39 |
| Bhadrakali Chok | 22 | 28.21 | 84.01 |
| Bhagi | 6 | 31.24 | 79.10 |
| Bhagirathi | 6 | 30.50 | 78.35 |
| Bhairav Tol | 22 | 28.23 | 83.99 |
| Bhaktapur | 6, 13 | 27.65 | 85.47 |
| Bhatbhateni | 20 | 53.77 | 89.10 |
| Bhattichok | 26 | 26.41 | 87.27 |
| Bheri | 6, 13 | 28.63 | 82.64 |
| Bhimkali Patan | 22 | 28.24 | 84.00 |
| Bhimsen Tol | 22 | 28.23 | 83.99 |
| Bhobi Khola | 20 | 53.77 | 89.12 |
| Bhojpur | 7, 13 | 27.18 | 87.05 |

| Name | Page | °N Latitude | °E Longitude |
|---|---|---|---|
| Bhote Bahal | 20 | 53.75 | 89.06 |
| Bhote Chautara | 22 | 28.18 | 84.03 |
| Bhratang | 178 | 28.58 | 84.19 |
| Bhrikuti Nager | 26 | 28.05 | 81.62 |
| Bhujung | 178 | 28.31 | 84.28 |
| Bichari Chok | 22 | 28.22 | 83.98 |
| Bijayapur Khola | 22 | 28.26 | 83.99 |
| Bilaspur | 6, 10 | 31.33 | 76.77 |
| Bindhyabasini | 22 | 28.24 | 83.99 |
| Bir Hospital, Kathmandu | 20 | 53.75 | 89.07 |
| Biratnagar | 7, 13 | 26.50 | 87.28 |
| Birauta | 22 | 28.18 | 84.03 |
| Birganj | 6, 13 | 27.02 | 84.87 |
| Bisa Nagar | 20 | 53.77 | 89.10 |
| Bisankh | 20 | 53.66 | 89.15 |
| Bishnumati River | 20 | 53.79 | 89.07 |
| Black Mountains | 7, 14 | 27.26 | 90.33 |
| Bode | 20 | 53.73 | 89.19 |
| Bomdila | 7, 14 | 27.43 | 92.16 |
| Bondey | 23 | 27.40 | 89.41 |
| Boudha | 20 | 53.77 | 89.13 |
| Brahmaputra | 7, 14, 26 | 26.45 | 87.26 |
| Brammah | 179 | 33.50 | 76.05 |
| Buddha Chok | 22 | 28.21 | 84.00 |
| Budhanilkantha | 20 | 53.82 | 89.15 |
| Bulbuliya | 26 | 28.05 | 81.63 |
| Buli | 23 | 27.48 | 90.69 |
| Bumrang | 7, 14 | 27.28 | 90.91 |
| Bumthang | 7, 14 | 27.52 | 90.78 |
| Bungmati | 20 | 53.68 | 89.04 |
| Butwal | 6, 13 | 27.71 | 83.46 |
| Chabahil | 20 | 53.76 | 89.14 |
| Chainpur | 6, 10, 13 | 29.55 | 81.20 |
| Chakhar | 23 | 27.53 | 90.74 |
| Chamba | 6, 10 | 32.55 | 76.14 |
| Chame | 178 | 28.55 | 84.26 |
| Chamoli | 6, 10 | 30.38 | 79.34 |
| Chandigarh | 6, 10 | 30.67 | 76.73 |
| Changu Narayan | 20 | 53.75 | 89.24 |
| Chapagaon | 20 | 53.64 | 89.07 |
| Charikot | 7, 13 | 27.68 | 86.00 |
| Chashma Shahi | 21 | 34.08 | 74.89 |
| Chauthe | 22 | 28.19 | 84.03 |
| Chelmstod Club | 25 | 31.10 | 77.17 |
| Chenab | 6, 10, 21 | 32.39 | 73.74 |
| Chhauni | 20 | 53.76 | 89.03 |
| Chhetrapati | 20 | 53.76 | 89.05 |
| Chhinedada | 22 | 28.18 | 84.01 |
| Chikanmugal | 20 | 53.75 | 89.05 |
| Chilas | 6, 10 | 35.42 | 74.09 |
| Chiple Dhunga | 22 | 28.22 | 83.99 |
| Chitwan | 6, 13 | 27.47 | 84.50 |
| Cho Oyu | 173 | 28.10 | 86.66 |
| Choedra | 23 | 27.49 | 90.71 |
| Chola La | 179 | 27.94 | 86.75 |
| Cholera Hospital, Kathmandu | 20 | 53.74 | 89.05 |
| Chomo Lhan | 7, 14, 13 | 27.82 | 89.28 |
| Chorten | 25 | 27.33 | 88.62 |
| Christ Church, Simla | 25 | 31.10 | 77.17 |
| Chumbi Valley | 7, 13, 14 | 27.92 | 89.21 |
| Chumophug | 23 | 27.49 | 89.39 |
| Chureta Patan | 22 | 28.21 | 84.02 |
| Churia Range | 6, 7, 10, 13 | 29.68 | 79.22 |
| Chuzom | 7, 14 | 27.31 | 89.56 |
| Community Center, Gangtok | 25 | 27.33 | 88.62 |
| Dachigam | 21 | 34.11 | 74.95 |
| Dadeldhura | 6, 10, 13 | 29.28 | 80.59 |
| Daduwa Kharka | 22 | 28.19 | 84.02 |
| Dailekh | 6, 10, 13 | 28.98 | 81.66 |
| Daji | 23 | 27.50 | 89.70 |
| Dakpathar | 25 | 30.38 | 78.07 |
| Dakshin Barahi | 20 | 53.71 | 89.17 |
| Dakshinkali | 20 | 53.64 | 88.99 |
| Dal Lake | 21 | 34.14 | 74.88 |
| Dallu | 20 | 53.76 | 89.05 |
| Damabandi | 26 | 26.47 | 87.25 |
| Damauli | 6, 10, 13 | 28.01 | 84.27 |
| Dana | 178 | 28.54 | 83.65 |
| Danagyu | 178 | 28.53 | 84.32 |

Map Index 203

| Name | Page | °N Latitude | °E Longitude |
|---|---|---|---|
| Dangel | 179 | 33.40 | 76.37 |
| Daporijo | 7, 14 | 28.00 | 94.26 |
| Darahia | 26 | 26.42 | 87.26 |
| Darcha | 179 | 32.65 | 77.21 |
| Darchula | 6, 10, 13 | 29.86 | 80.63 |
| Darjeeling | 7, 13 | 27.04 | 88.26 |
| David Falls | 22 | 28.19 | 83.96 |
| Decheng | 23 | 27.50 | 90.80 |
| Deer Park, Gangtok | 25 | 27.33 | 88.62 |
| Dehra Dun | 6, 10 | 30.38 | 78.07 |
| Delhi | 6, 10, | 28.52 | 77.22 |
| Deosai Plains | 6, 10 | 34.90 | 75.42 |
| Deurali | 22 | 28.18 | 84.03 |
| Deurali Chok | 22 | 28.18 | 84.02 |
| Devkotachok | 26 | 26.45 | 87.29 |
| Dhading | 6, 13 | 27.89 | 84.85 |
| Dhaka | 7 | 23.71 | 90.39 |
| Dhampu | 178 | 28.65 | 83.60 |
| Dhankuta | 7, 13 | 26.99 | 87.34 |
| Dharapani | 178 | 28.52 | 84.36 |
| Dharmashalatol | 26 | 26.49 | 87.29 |
| Dharmsala | 6, 10 | 32.17 | 76.31 |
| Dhaulagiri | 6, 10, 13 | 28.73 | 83.32 |
| Dhiprang | 178 | 28.34 | 83.96 |
| Dhomboji | 26 | 28.07 | 81.63 |
| Dhorpatan | 6, 10, 13 | 28.50 | 83.06 |
| Dhumre | 22 | 28.18 | 84.02 |
| Dhunche | 6, 13 | 28.09 | 85.30 |
| Dhunge Sangu | 22 | 28.21 | 84.02 |
| Dibrugarh | 7, 14 | 27.40 | 94.92 |
| Dihiko Patan | 22 | 28.21 | 83.96 |
| Dingboche | 179 | 27.87 | 86.81 |
| Dipendra Shabhagriha | 22 | 28.21 | 83.99 |
| Dolpo | 6, 10, 13 | 29.05 | 83.21 |
| Domkhar Tashichoeling | 23 | 27.45 | 90.69 |
| Drangyekha | 23 | 27.44 | 89.37 |
| Drela Dzong | 23 | 27.43 | 89.44 |
| Druk Choeding | 23 | 27.43 | 89.39 |
| Drukyel Dzong | 23 | 27.49 | 89.33 |
| Dudhkosi | 7, 13 | 27.41 | 86.69 |
| Dunai | 6, 10, 13 | 28.93 | 82.91 |
| Dungtse Lhakhang | 23 | 27.44 | 89.40 |
| Durganath Mandir | 21 | 34.07 | 74.83 |
| Dzongdrakha | 23 | 27.39 | 89.41 |
| Ektobhatti | 178 | 28.83 | 83.78 |
| Enchery Monastery, Gangtok | 25 | 27.33 | 88.62 |
| Floating Garden, Srinagar | 21 | 34.13 | 74.84 |
| Forest Research Institute, Dehra Dun | 25 | 30.38 | 78.07 |
| Gahelitol | 26 | 26.45 | 87.29 |
| Gairha Patan | 22 | 28.23 | 83.99 |
| Gairidhara | 20 | 53.77 | 89.09 |
| Gamgadhi | 6, 10, 13 | 29.57 | 82.26 |
| Ganda La | 179 | 34.01 | 77.43 |
| Ganderbal | 21 | 34.21 | 74.77 |
| Gandhi Park, Srinagar | 21 | 34.07 | 74.79 |
| Ganesh Tol | 22 | 28.23 | 83.98 |
| Ganeshpur | 26 | 28.06 | 81.63 |
| Ganga Path | 20 | 53.75 | 89.06 |
| Ganges | 6, 7, 13, 14 | 25.31 | 86.18 |
| Gangtok | 7, 13, 14 | 27.33 | 88.62 |
| Gantey | 23 | 27.42 | 89.39 |
| Garhi | 25 | 30.38 | 78.07 |
| Garhwal | 6, 10 | 30.50 | 79.09 |
| Garhwal Mandal Vikas Nigam | 25 | 30.38 | 78.07 |
| Gasa | 7, 14 | 27.87 | 89.74 |
| Gasherbrum | 6, 10 | 35.66 | 76.88 |
| Gaukharka | 22 | 28.16 | 84.04 |
| Gaur | 6, 13 | 26.86 | 85.30 |
| Gauri Sankar | 7, 13 | 27.97 | 86.33 |
| Gausala | 20 | 53.75 | 89.11 |
| Geylegphug | 7, 14 | 26.85 | 90.45 |
| Ghandruk | 178 | 28.38 | 83.82 |
| Gharbari | 26 | 28.06 | 81.63 |
| Ghari Patan | 22 | 28.19 | 83.98 |
| Ghasa | 178 | 28.61 | 83.64 |
| Ghorepani | 178 | 28.41 | 83.71 |
| Gilgit | 6, 10 | 35.92 | 74.31 |
| Godavari | 20 | 53.64 | 89.14 |

| Name | Page | °N Latitude | °E Longitude |
|---|---|---|---|
| Gokameswor | 20 | 53.78 | 89.18 |
| Gokarna Rain Forest | 20 | 53.77 | 89.19 |
| Gorakhpur | 6, 13, | 26.73 | 83.52 |
| Gordon Castle, Simla | 25 | 31.10 | 77.17 |
| Gorina | 23 | 27.42 | 89.38 |
| Gorkha | 6, 13 | 27.98 | 84.59 |
| Gosaigaun | 26 | 28.04 | 81.62 |
| Guheswari | 20 | 53.76 | 89.12 |
| Gullol Gali La | 21 | 32.63 | 74.37 |
| Gurudwara | 25 | 31.10 | 77.17 |
| Guru Ramrai Darbar | 25 | 30.38 | 78.07 |
| Gyalung | 23 | 27.40 | 89.71 |
| Gyarjati | 22 | 28.24 | 83.97 |
| Gyetsa | 23 | 27.47 | 90.68 |
| | | | |
| Hadigaon | 20 | 53.76 | 89.10 |
| Hari Parbat Fort, Srinagar | 21 | 34.11 | 74.80 |
| Hariyo Kharka | 22 | 28.19 | 84.00 |
| Harshidh | 20 | 53.67 | 89.10 |
| Hatkhola | 26 | 26.45 | 87.29 |
| Hazratbal Mosque, Srinagar | 21 | 34.15 | 74.84 |
| Hemis | 6, 10 | 33.91 | 77.72 |
| Hetauda | 6, 13 | 27.42 | 85.05 |
| Heyphug | 23 | 27.39 | 89.45 |
| Himachal | 25 | 31.10 | 77.17 |
| Himala Tol | 22 | 28.20 | 83.98 |
| Hinku Cave | 178 | 28.50 | 83.89 |
| Hospital Chok, Pokhara | 22 | 28.21 | 84.00 |
| Huang He | 7 | 39.67 | 106.79 |
| Humde | 178 | 28.64 | 84.10 |
| Humla | 6, 10, 13 | 30.12 | 81.78 |
| | | | |
| Ichangu Narayan | 20 | 53.78 | 89.00 |
| Idgah | 21 | 34.11 | 74.77 |
| Ikarahi | 26 | 26.48 | 87.26 |
| Ilam | 7, 13 | 26.93 | 87.91 |
| Indrachowk | 20 | 53.75 | 89.06 |
| Indus | 6 | 33.61 | 78.07 |
| Irrawaddy | 7 | 26.28 | 98.26 |
| Islamabad | 6, 10, | 33.72 | 73.07 |
| Island Peak | 179 | 27.90 | 86.95 |
| Itanagar | 7, 14 | 27.00 | 93.59 |
| | | | |
| Jajarkot | 6, 10, 13 | 28.70 | 82.22 |
| Jakar Dzong | 23 | 27.51 | 90.74 |
| Jakar Lhakhang | 23 | 27.50 | 90.75 |
| Jalpaiguri | 7, 13 | 26.52 | 88.65 |
| Jami Masjid | 21 | 34.10 | 74.80 |
| Jammu | 6, 10 | 32.71 | 74.85 |
| Jampey Lhakhang | 23 | 27.53 | 90.73 |
| Jamungachhitol | 26 | 26.47 | 87.27 |
| Janakpur | 7, 13 | 26.73 | 85.92 |
| Janakpuria Chok | 22 | 28.20 | 84.01 |
| Janapathtol | 26 | 26.44 | 87.27 |
| Jara La | 6, 10 | 32.87 | 79.45 |
| Jarebar | 22 | 28.21 | 83.97 |
| Jatuwa | 26 | 26.43 | 87.29 |
| Jawahar Tunner | 21 | 32.63 | 74.37 |
| Jawaharlal Nehru Memorial Park | 21 | 34.10 | 74.89 |
| Jhapa | 7, 13 | 26.63 | 88.08 |
| Jharkot | 178 | 28.82 | 83.84 |
| Jiley La | 23 | 27.38 | 89.34 |
| Jomsom | 6, 10, 13 | 28.78 | 83.71 |
| Jorpati | 20 | 53.76 | 89.16 |
| Joshimath | 6, 10, 13 | 30.56 | 79.56 |
| Jumla | 6, 10, 13 | 29.23 | 82.13 |
| Junction Kharka | 178 | 28.37 | 83.93 |
| Jungshina | 23 | 27.46 | 89.67 |
| Jwalatole | 26 | 26.49 | 87.26 |
| | | | |
| K2 | 6, 10 | 35.88 | 76.51 |
| Kabutar Khan | 21 | 34.11 | 74.85 |
| Kagbeni | 178 | 28.84 | 83.80 |
| Kahukhola Gaon | 22 | 28.22 | 84.02 |
| Kahun Danda | 22 | 28.24 | 84.01 |
| Kaji Pokhari | 22 | 28.19 | 84.01 |
| Kakarvitta | 7, 13 | 26.63 | 88.08 |
| Kalaktang | 7, 14 | 27.11 | 92.12 |
| Kalanga Fort | 25 | 30.38 | 78.07 |
| Kalapani | 6, 10, 13 | 30.12 | 80.97 |
| Kala Pattar | 179 | 28.01 | 86.77 |

| Name | Page | °N Latitude | °E Longitude |
|---|---|---|---|
| Kaldhara | 20 | 53.76 | 89.05 |
| Kali Barl Mandir | 25 | 31.10 | 77.17 |
| Kaligandaki | 6, 13 | 28.35 | 83.57 |
| Kalimpong | 7, 13, 14 | 27.07 | 88.48 |
| Kalopani | 178 | 28.65 | 83.60 |
| Kalsi | 25 | 30.38 | 78.07 |
| Kamal Pokhari | 22 | 28.22 | 84.02 |
| Kanchanbari | 26 | 26.48 | 87.28 |
| Kanchenjunga | 7, 13 | 27.71 | 88.06 |
| Kangkar Tesi (Kailas) | 6, 10 | 31.13 | 81.22 |
| Kangra Valley | 6, 10 | 31.87 | 76.62 |
| Kangto | 7, 14 | 27.90 | 92.55 |
| Karakoram Pass | 6, 10 | 35.48 | 77.75 |
| Karakoram Range | 6, 10 | 36.19 | 75.41 |
| Karcha gompa | 179 | 33.52 | 76.90 |
| Kargil | 6, 10 | 34.54 | 76.14 |
| Karkhanatol | 26 | 26.46 | 87.26 |
| Karnali | 6, 13 | 29.21 | 82.05 |
| Kasar | 179 | 33.51 | 76.83 |
| Kathkuppa | 26 | 26.41 | 87.29 |
| Kathmandu | 6, 13, 20 | 27.70 | 85.32 |
| Kati | 6, 13 | 30.06 | 80.28 |
| Katunje | 20 | 53.70 | 89.20 |
| Kaza | 6, 10 | 31.91 | 78.28 |
| Kepsang La | 6, 10 | 34.04 | 79.37 |
| Khahare | 22 | 28.22 | 83.96 |
| Khandbari | 7, 13 | 27.47 | 87.25 |
| Khankanpurwa | 26 | 28.04 | 81.62 |
| Kharjitol | 26 | 26.49 | 87.28 |
| Khatkanpurwa | 26 | 28.05 | 81.63 |
| Khatung Kang | 178 | 28.78 | 83.91 |
| Khingar | 178 | 28.83 | 83.82 |
| Khokana | 20 | 53.68 | 89.03 |
| Khokshaha | 26 | 26.38 | 87.27 |
| Khumjung | 179 | 27.82 | 86.71 |
| Khundel | 179 | 33.36 | 76.23 |
| Khusar Sar | 21 | 34.13 | 74.78 |
| Kibai | 179 | 32.33 | 78.11 |
| Kibbur | 6, 10 | 32.33 | 78.00 |
| Kilar | 6, 10, 21 | 33.05 | 76.42 |
| Kimdol | 20 | 53.76 | 89.02 |
| Kirtipur | 20 | 53.73 | 89.00 |

| Name | Page | °N Latitude | °E Longitude |
|---|---|---|---|
| Kishtwar | 6, 10, 21 | 33.27 | 75.76 |
| Kochakhaltol | 26 | 26.46 | 87.27 |
| Kodari | 7, 13 | 27.89 | 85.92 |
| Kohiti | 20 | 53.75 | 89.05 |
| Kohobara Nayatole | 26 | 26.50 | 87.27 |
| Kolahol Glacier | 21 | 34.14 | 75.34 |
| Konchongsum | 23 | 27.53 | 90.75 |
| Koriyanpur | 26 | 28.06 | 81.62 |
| Koteshwore | 20 | 53.73 | 89.11 |
| Krishna Mandir Chok | 22 | 28.20 | 83.99 |
| Kula Kangri | 7, 14 | 28.22 | 90.63 |
| Kuloko Dil | 22 | 28.22 | 83.97 |
| Kulu | 6, 10 | 31.98 | 77.11 |
| Kumaon | 6, 10, 13 | 29.74 | 79.68 |
| Kun | 179 | 34.01 | 76.07 |
| Kungachoeling | 23 | 27.46 | 89.36 |
| Kungi La | 179 | 34.10 | 76.46 |
| Kungzandra | 23 | 27.52 | 90.80 |
| Kurjey | 23 | 27.54 | 90.73 |
| Kuwar Tol | 22 | 28.22 | 83.98 |
| Kyampin Chok | 22 | 28.21 | 83.96 |
| Kye Gompa | 179 | 32.28 | 78.07 |
| Kyelang | 6, 10 | 32.54 | 77.01 |
| Kyichu Lhakhang | 23 | 27.45 | 89.35 |
| Ladakh | 6, 10 | 34.68 | 77.91 |
| Lahaul Valley | 6, 10 | 32.35 | 77.26 |
| Lahore | 6, 10, | 31.53 | 74.34 |
| Lakeside, Pokhara | 22 | 28.21 | 83.96 |
| Laligurans | 22 | 28.21 | 84.01 |
| Lall Market, Gangtok | 25 | 27.33 | 88.62 |
| Lalten Bazar, Pokhara | 22 | 28.25 | 83.99 |
| Lamayuru | 6, 10 | 34.25 | 76.79 |
| Lamey Gompa | 23 | 27.50 | 90.73 |
| Lampata | 178 | 28.33 | 84.40 |
| Lamtara | 22 | 28.17 | 84.00 |
| Langtang | 6, 13 | 28.10 | 85.65 |
| Larjung | 178 | 28.69 | 83.61 |
| Laxman Sidh | 25 | 30.38 | 78.07 |
| Leh | 6, 10, 179 | 34.18 | 77.58 |
| Leinchour | 20 | 53.77 | 89.07 |
| Letdar | 178 | 28.75 | 83.97 |

| Name | Page | °N Latitude | °E Longitude |
|---|---|---|---|
| Lhasa | 7, 14 | 29.66 | 91.14 |
| Lhedang | 7, 14 | 27.91 | 90.67 |
| Lhotse | 179 | 27.96 | 86.95 |
| Lhuntshi | 7, 14 | 27.65 | 91.20 |
| Liddarwal | 21 | 34.11 | 75.22 |
| Lobuje | 179 | 27.96 | 86.78 |
| Lubhu | 20 | 53.68 | 89.14 |
| Lukha | 179 | 27.67 | 86.74 |
| Lukla | 7, 13 | 27.70 | 86.71 |
| | | | |
| Machepuchare Base Camp | 178 | 28.53 | 83.91 |
| Machhegaon | 20 | 53.71 | 88.97 |
| Mahabharat Range | 6, 10, 13 | 28.26 | 83.06 |
| Mahandranagar | 6, 10, 13 | 29.02 | 80.21 |
| Mahatgauda | 22 | 28.18 | 83.99 |
| Mahendrachok | 26 | 26.45 | 87.28 |
| Mahendra Pul | 22 | 28.22 | 84.00 |
| Majheri Patan | 22 | 28.18 | 84.02 |
| Majuwa | 22 | 28.19 | 84.00 |
| Makalu | 7, 13 | 27.90 | 87.10 |
| Makhar | 23 | 27.46 | 90.75 |
| Male Patan | 22 | 28.22 | 83.98 |
| Malhanawa | 26 | 26.43 | 87.26 |
| Malintan | 7, 14 | 28.09 | 92.97 |
| Malsi Deer Park | 25 | 30.38 | 78.07 |
| Manali | 179 | 32.10 | 77.13 |
| Manang | 178 | 28.67 | 84.01 |
| Manas | 7, 14 | 26.81 | 90.91 |
| Manaslu | 6, 13 | 28.58 | 84.55 |
| Mandi | 6, 10 | 31.70 | 76.94 |
| Mangadh | 26 | 26.50 | 87.29 |
| Mangan | 7, 13 | 27.33 | 88.50 |
| Manshera | 6, 10 | 34.34 | 73.20 |
| Mara | 7, 14 | 28.27 | 94.07 |
| Mardi Himal | 178 | 28.47 | 83.94 |
| Mardi Himal Base Camp | 178 | 28.45 | 83.94 |
| Marpha | 178 | 28.75 | 83.69 |
| Martyr's Gate | 20 | 53.75 | 89.07 |
| Masbar | 22 | 28.21 | 83.98 |
| Masina | 22 | 28.17 | 84.02 |
| Masinotara | 22 | 28.18 | 83.98 |
| Masu | 179 | 33.32 | 76.19 |
| Mate | 22 | 28.17 | 84.03 |
| Materwa | 26 | 26.39 | 87.27 |
| Matipani | 22 | 28.22 | 84.01 |
| Mekong | 7, 14 | 31.28 | 97.16 |
| Miging | 7, 14 | 28.84 | 94.73 |
| Milan Chok | 22 | 28.20 | 84.00 |
| Mipi | 7, 14 | 28.95 | 95.80 |
| Miruwa | 22 | 28.24 | 83.99 |
| Miruwa Chok | 22 | 28.24 | 83.98 |
| Mishmi Hills | 7, 14 | 28.49 | 96.57 |
| Mohariya Tol | 22 | 28.24 | 83.99 |
| Mongar | 7, 14 | 27.25 | 91.19 |
| Monjo | 179 | 27.77 | 86.73 |
| Mt. Everest (Sagarmatha) | 7, 13, 179 | 27.96 | 86.93 |
| Mugu | 6, 13 | 29.57 | 82.26 |
| Muktinath | 6, 10, 13 | 28.82 | 83.87 |
| Multhok | 22 | 28.21 | 83.97 |
| Mussoorie | 6, 10 | 30.52 | 78.06 |
| Mustan Chok | 22 | 28.20 | 83.98 |
| Muzaffarabad | 6, 10 | 34.35 | 73.48 |
| | | | |
| Nachani | 6, 10, 13 | 29.88 | 80.15 |
| Nagarkot | 20 | 53.75 | 89.37 |
| Nagdhunga | 22 | 28.20 | 83.98 |
| Nagin Lake | 21 | 34.13 | 74.82 |
| Nainital | 6, 10 | 29.40 | 79.44 |
| Najaregaon | 178 | 28.25 | 84.32 |
| Nakadesh | 20 | 53.73 | 89.17 |
| Nakuhu | 20 | 53.71 | 89.04 |
| Nalamukh | 22 | 28.23 | 83.99 |
| Namcha | 7, 13 | 27.90 | 86.70 |
| Namche Bazaar | 179 | 27.81 | 86.70 |
| Nanda Devi | 6, 10, 13 | 30.38 | 79.97 |
| Nanga Parbat | 6, 10 | 35.24 | 74.60 |
| Naradevi Pyabphal | 20 | 53.75 | 89.05 |
| Narayangarh | 6, 13 | 27.66 | 84.40 |
| Narayani | 6, 13 | 27.95 | 84.18 |
| National Museum, Kathmandu | 20 | 53.75 | 89.03 |
| Naudanda | 178 | 28.29 | 83.86 |

| Name | Page | °N Latitude | °E Longitude |
|---|---|---|---|
| Naya Bazar, Pokhara | 22 | 28.22 | 83.99 |
| Naya Pul | 178 | 28.30 | 83.78 |
| Nayagaon | 22 | 28.20 | 83.99 |
| Nehru Park, Srinagar | 21 | 34.10 | 74.84 |
| Nepalganj | 6, 10, 13, 26 | 28.06 | 81.62 |
| New Road | 22 | 28.22 | 83.99 |
| Ngang Lhatkang | 23 | 27.59 | 90.74 |
| Nichnai La | 21 | 32.63 | 74.37 |
| North Lakhimpu | 7, 14 | 27.29 | 94.13 |
| Nun | 6, 10, 179 | 33.99 | 76.02 |
| Nuniatol | 26 | 26.42 | 87.27 |
| Nuptse | 179 | 27.96 | 86.88 |
| Nyimalung | 23 | 27.46 | 90.75 |
| Nyuru | 179 | 33.28 | 77.04 |
|  |  |  |  |
| Oberoi Palace, Srinagar | 21 | 34.08 | 74.87 |
| Okhaldhunga | 7, 13 | 27.31 | 86.50 |
| Orchid Sanctuary, Gangtok | 25 | 27.33 | 88.62 |
|  |  |  |  |
| Padam | 6, 10 | 33.40 | 76.90 |
| Pahalgam | 21 | 33.98 | 75.30 |
| Paknajol | 20 | 53.77 | 89.06 |
| Panga Chovar | 20 | 53.72 | 89.01 |
| Pangbeysa | 23 | 27.36 | 89.42 |
| Pangboche | 179 | 27.85 | 86.78 |
| Pangdu | 23 | 27.41 | 89.70 |
| Pangri | 23 | 27.49 | 89.68 |
| Pardi Bazaar, Pokhara | 22 | 28.20 | 83.98 |
| Pardi Dam | 22 | 28.20 | 83.97 |
| Pardi Khola | 22 | 28.19 | 83.97 |
| Pari Mahal | 21 | 34.08 | 74.88 |
| Paro | 7, 13, 14 | 27.44 | 89.42 |
| Paroik Pass | 7 | 36.84 | 75.42 |
| Pashupati Temple, Kathmandu | 20 | 53.75 | 89.12 |
| Pasighat | 7, 14 | 28.04 | 95.33 |
| Patan | 6, 13, 22 | 27.63 | 85.36 |
| Patan Besi | 20 | 53.72 | 89.08 |
| Pathankot | 6, 10 | 32.18 | 75.58 |
| Pathar Mosque, Srinagar | 21 | 34.09 | 74.79 |

| Name | Page | °N Latitude | °E Longitude |
|---|---|---|---|
| Pemagatsel | 7, 14 | 27.01 | 91.37 |
| Phakding | 179 | 27.72 | 86.71 |
| Phale Patan | 22 | 28.19 | 84.00 |
| Pharping | 20 | 53.65 | 88.99 |
| Phericho | 179 | 27.86 | 86.80 |
| Phewa Tal | 22 | 28.22 | 83.95 |
| Phuktal gompa | 179 | 33.27 | 77.17 |
| Phulbari | 22 | 28.23 | 84.00 |
| Phultakra | 26 | 28.06 | 81.64 |
| Phuntsholing | 7, 14 | 26.87 | 89.39 |
| Picharatol | 26 | 26.45 | 87.27 |
| Pirlipurwa | 26 | 28.04 | 81.63 |
| Pishu | 179 | 33.61 | 77.00 |
| Pithoragarh | 6, 10, 13 | 29.59 | 80.21 |
| Pokangnang | 23 | 27.47 | 89.68 |
| Pokhara | 6, 10, 13, 22, 26 | 28.17 | 83.99 |
| Prasyan | 22 | 28.22 | 83.98 |
| Pratinagar | 22 | 28.20 | 83.99 |
| Prithvi Chok | 22 | 28.21 | 83.99 |
| Punch | 6, 10, 21 | 33.78 | 74.09 |
| Purano Dhara | 22 | 28.22 | 84.00 |
| Puspalalchok | 26 | 26.49 | 87.29 |
| Pyuthan | 6, 10, 13 | 28.29 | 82.79 |
|  |  |  |  |
| Rabi Bhawan | 20 | 53.74 | 89.03 |
| Raghunath Mandir | 21 | 34.08 | 74.79 |
| Rajbansitol | 26 | 26.42 | 87.30 |
| Raj Bhatvan (Peterhoff) | 25 | 31.10 | 77.17 |
| Raj Bhavan | 21 | 34.09 | 74.89 |
| Rakaposhi | 6, 10 | 36.14 | 74.49 |
| Ramakrishna Ashram | 25 | 30.38 | 78.07 |
| Ram Bazar, Pokhara | 22 | 28.20 | 84.00 |
| Rambong Pass | 178 | 28.44 | 84.27 |
| Ramechap | 7, 13 | 27.37 | 86.00 |
| Ram Ghat | 22 | 28.22 | 84.00 |
| Ramkrishna Chok | 22 | 28.23 | 83.98 |
| Ramnagar | 6, 10 | 29.39 | 79.13 |
| Ram Nager | 26 | 28.06 | 81.64 |
| Rampur | 6, 10 | 31.45 | 77.64 |
| Ramtopla | 23 | 27.37 | 89.68 |
| Rangdom Gompa | 6, 10 | 34.00 | 76.39 |
| Rani | 26 | 26.40 | 87.27 |

| Name | Page | °N Latitude | °E Longitude |
|---|---|---|---|
| Rani Ban | 22 | 28.19 | 83.99 |
| Ranipauwa | 22 | 28.22 | 84.00 |
| Rani Pokhari | 26 | 26.40 | 87.28 |
| Ratna Chok | 22 | 28.21 | 83.98 |
| Ratna Mandir Palace, Pokhara | 22 | 28.21 | 83.96 |
| Rato Pairha | 22 | 28.18 | 83.99 |
| Redshestol | 26 | 26.44 | 87.28 |
| Riga | 7, 14 | 28.39 | 95.06 |
| Rinpung Dzong | 23 | 27.43 | 89.41 |
| Rishikesh | 6, 10 | 30.07 | 78.26 |
| Robbers Cave, Dehra Dun | 25 | 30.38 | 78.07 |
| Rohatepani | 22 | 28.25 | 83.98 |
| Ropri Gali La | 21 | 32.63 | 74.37 |
| Royal Palace, Kathmandu | 20 | 53.76 | 89.07 |
| Rozahbal Mosque, Srinagar | 21 | 34.10 | 74.81 |
| Rukse Chahara | 178 | 28.55 | 83.64 |
| Rumbak | 179 | 34.02 | 77.52 |
| Rusi La | 179 | 34.29 | 76.15 |
| Saharanpur | 6, 10 | 29.96 | 77.51 |
| Saharanpur Chowk | 25 | 30.38 | 78.07 |
| Sahastra-Dhara | 25 | 30.38 | 78.07 |
| Salleri | 7, 13 | 27.49 | 86.74 |
| Salween | 7, 14 | 30.48 | 96.51 |
| Salyan | 6, 10, 13 | 28.34 | 82.12 |
| Salyanibag | 26 | 28.06 | 81.63 |
| Samdrup | 7, 14 | 26.90 | 91.44 |
| Samtenling | 23 | 27.47 | 90.69 |
| Sanepa | 20 | 53.73 | 89.05 |
| Sanga | 20 | 53.67 | 89.31 |
| Sangbay | 7, 13, 14 | 27.05 | 89.18 |
| Sangla | 6, 10 | 31.42 | 78.33 |
| Sangnag Choekhor | 23 | 27.45 | 89.37 |
| Sankhu | 20 | 53.76 | 89.29 |
| Sankosh | 7, 14 | 27.79 | 90.06 |
| Sanogaon | 20 | 53.69 | 89.12 |
| Sarangkot | 22, 178 | 28.25 | 83.98 |
| Saraswatitol | 26 | 26.46 | 87.29 |
| Sarbhang | 7, 14 | 26.87 | 90.25 |
| Sarchu | 6, 10 | 32.81 | 77.48 |
| Sarkhang | 7, 14 | 26.86 | 90.21 |
| Satsam Chorten | 23 | 27.47 | 89.34 |
| Sengi la | 179 | 33.97 | 76.78 |
| Sentmeri Chok | 22 | 28.20 | 84.00 |
| Seppa | 7, 14 | 27.29 | 93.00 |
| Seri Gandaki | 22 | 28.19 | 83.99 |
| Sesha Narayan | 20 | 53.66 | 88.99 |
| Seti Gandaki River | 22 | 28.25 | 83.98 |
| Shahid Chok | 22 | 28.20 | 83.97 |
| Shankar Tol | 22 | 28.24 | 83.99 |
| Shanshahi Ashram | 25 | 30.38 | 78.07 |
| Shantichok | 26 | 26.45 | 87.27 |
| Shantihat | 26 | 26.46 | 87.28 |
| Shanti Nagar Chok | 22 | 28.21 | 83.97 |
| Shanti Tole | 22 | 28.18 | 84.01 |
| Shemgang | 7, 14 | 27.19 | 90.63 |
| Shey Gompa | 6, 10, 13 | 29.35 | 82.82 |
| Shigatse | 7, 13 | 29.14 | 88.95 |
| Shing-Kun La | 179 | 32.77 | 77.15 |
| Shisapangma | 6, 13 | 28.41 | 85.59 |
| Shivalaya | 22 | 28.16 | 84.01 |
| Shya Gang | 178 | 28.77 | 83.94 |
| Siddharthachok | 26 | 26.45 | 87.28 |
| Sikha | 178 | 28.44 | 83.68 |
| Sikkim | 7, 13 | 27.58 | 88.40 |
| Silgadhi | 6, 10, 13 | 29.26 | 80.91 |
| Siliguri | 7, 13 | 26.71 | 88.36 |
| Simal Chok | 22 | 28.21 | 83.98 |
| Simikot | 6, 10, 13 | 29.96 | 81.85 |
| Simla | 6, 10, 25 | 31.10 | 77.17 |
| Simpani | 22 | 28.24 | 83.98 |
| Sindhuli | 7, 13 | 26.69 | 86.33 |
| Singh Durbar | 20 | 53.74 | 89.08 |
| Sintoka Dzong | 23 | 27.39 | 89.72 |
| Siromanichok | 26 | 26.46 | 87.27 |
| Sirsir la | 179 | 33.99 | 76.75 |
| Sitaghat | 22 | 28.18 | 84.00 |
| Sitaladevi Mandir | 22 | 28.22 | 84.01 |
| Sitapaila | 22 | 28.17 | 83.99 |
| Skardu | 6, 10 | 35.31 | 75.54 |
| Skyumpata | 179 | 33.91 | 76.75 |

| Name | Page | °N Latitude | °E Longitude |
|---|---|---|---|
| Solan | 6, 10 | 30.99 | 77.12 |
| Solukhumbu | 7, 13 | 27.76 | 86.60 |
| Sombrang | 23 | 27.46 | 90.86 |
| Sonamarg | 6, 10 | 34.29 | 75.29 |
| Spadum | 179 | 33.42 | 76.89 |
| Spiti Valley | 6, 10 | 31.45 | 78.54 |
| Srinagar | 6, 10 | 34.06 | 74.80 |
| Suban | 7, 14 | 27.76 | 94.53 |
| Subansiri | 7, 14 | 27.05 | 94.16 |
| Suikhet | 178 | 28.29 | 83.91 |
| Sundarijal | 20 | 53.79 | 89.23 |
| Sunkoth | 20 | 53.68 | 89.06 |
| Surkhet | 6, 10, 13 | 28.60 | 81.60 |
| Suru | 6, 10 | 34.16 | 75.94 |
| Suryabinayak | 20 | 53.69 | 89.22 |
| Sut | 6 | 31.54 | 78.45 |
| Sutlej | 6 | 31.41 | 77.63 |
| Swayambhu | 20 | 53.76 | 89.03 |
| Syaldule | 22 | 28.20 | 83.98 |
| Syani Patan | 22 | 28.18 | 84.00 |
| Syanje | 178 | 28.43 | 84.39 |
| Syauli Bazaar | 178 | 28.35 | 83.81 |
| Syr Darya | 7 | 41.46 | 75.83 |
| | | | |
| Taba | 23 | 27.45 | 89.69 |
| Tadapani | 178 | 28.40 | 83.77 |
| Ta Dzong (National Museum) | 23 | 27.43 | 89.42 |
| Taghung | 178 | 28.42 | 83.82 |
| Taktsang | 23 | 27.48 | 89.36 |
| Tallo Dip | 22 | 28.25 | 83.99 |
| Tamar | 7, 13 | 26.92 | 87.20 |
| Tangsebi | 23 | 27.45 | 90.82 |
| Tangsey | 6, 10 | 34.02 | 78.19 |
| Tansen | 6, 13 | 27.87 | 83.55 |
| Tanze | 179 | 33.14 | 77.22 |
| Tapkeshwar | 25 | 30.38 | 78.07 |
| Taplejung | 7, 13 | 27.24 | 87.51 |
| Tarancha | 178 | 28.30 | 84.39 |
| Ta Rimochen | 23 | 27.54 | 90.85 |
| Tarke Kang (Glacier Dome) | 178 | 28.61 | 83.90 |

| Name | Page | °N Latitude | °E Longitude |
|---|---|---|---|
| Tar La | 179 | 34.25 | 76.92 |
| Tashi Chho Dzong | 23 | 27.44 | 89.68 |
| Tashigang | 6, 10 | 32.91 | 79.20 |
| Tashingang | 7, 14 | 27.28 | 91.45 |
| Tashing Village | 22 | 28.19 | 83.96 |
| Tashi Yangtse | 7, 14 | 27.57 | 91.51 |
| Tatopani | 178 | 28.49 | 83.65 |
| Taudaha | 20 | 53.69 | 89.01 |
| Tawang | 7, 14 | 27.58 | 91.88 |
| Tehri | 6, 10 | 30.45 | 78.47 |
| Teju | 7, 14 | 27.88 | 96.26 |
| Tengpocha | 179 | 27.83 | 86.77 |
| Tersapatti | 22 | 28.23 | 83.99 |
| Tezpur | 7, 14 | 26.68 | 92.80 |
| Thaiba | 20 | 53.66 | 89.11 |
| Thaleku | 178 | 28.56 | 84.22 |
| Thami | 179 | 27.86 | 86.62 |
| Thangbi | 23 | 27.56 | 90.72 |
| Thankot | 20 | 53.75 | 88.90 |
| Thanza | 7, 14 | 28.01 | 90.22 |
| Tharpaling | 23 | 27.49 | 90.70 |
| Thecho | 20 | 53.65 | 89.07 |
| Thimphu | 7, 14 | 27.44 | 89.67 |
| Thini | 178 | 28.78 | 83.72 |
| Thorung La | 178 | 28.79 | 83.93 |
| Thorung Phedi | 178 | 28.78 | 83.97 |
| Thowadra | 23 | 27.59 | 90.87 |
| Thunkar | 7, 14 | 27.81 | 91.12 |
| Tikathali | 20 | 53.70 | 89.12 |
| Tilicho Lake | 178 | 28.70 | 83.86 |
| Timang Besi | 178 | 28.53 | 84.31 |
| Tinpaini | 26 | 26.46 | 87.29 |
| Tintolia | 26 | 26.44 | 87.28 |
| Tirido | 6, 10 | 33.59 | 78.07 |
| Tirkhedhunga | 178 | 28.35 | 83.75 |
| Tista | 7, 13, 14 | 26.62 | 88.69 |
| Trongsa | 7, 14 | 27.50 | 90.50 |
| Tribhuvan International Airport | 20 | 53.74 | 89.13 |
| Trishuli | 6, 13 | 28.69 | 85.30 |
| Tsalimaphe | 23 | 27.40 | 89.70 |
| Tsangpo | 7, 13, 14 | 29.38 | 94.36 |

210  Map Index

| Name | Page | °N Latitude | °E Longitude |
|---|---|---|---|
| Tukuche | 178 | 28.72 | 83.66 |
| Tulsipur | 6, 10, 13 | 28.15 | 82.29 |
| Tundikhel | 22 | 28.24 | 83.99 |
| Tunga Pass | 7, 14 | 28.96 | 94.10 |
| Tutunga | 22 | 28.18 | 84.00 |
| Uchu | 23 | 27.41 | 89.40 |
| Udaipur | 6, 10 | 32.68 | 76.72 |
| Udhampur | 6, 10 | 33.07 | 75.31 |
| Ugyen Choeling | 23 | 27.55 | 90.85 |
| Ugyen Pelri Palace | 23 | 27.43 | 89.40 |
| Umasi La | 179 | 33.43 | 76.49 |
| Upallu Dip | 22 | 28.25 | 83.99 |
| Upper Pisang | 178 | 28.61 | 84.15 |
| Ura | 23 | 27.44 | 90.86 |
| Ura La | 23 | 27.44 | 90.84 |
| Usta | 178 | 28.31 | 84.40 |
| Uttarkashi | 6, 10 | 30.73 | 78.44 |

| Name | Page | °N Latitude | °E Longitude |
|---|---|---|---|
| Wangdi Che Monastery | 23 | 27.45 | 89.67 |
| Wangdi Phodrang | 7, 14 | 27.46 | 89.89 |
| Xi Jiang | 7 | 24.43 | 107.49 |
| Yak Kharka | 178 | 28.70 | 83.98 |
| Yamunotri | 6, 10 | 30.93 | 78.55 |
| Yangothang | 7, 13, 14 | 27.76 | 89.37 |
| Yangtze | 7 | 31.95 | 98.45 |
| Yengal | 20 | 53.75 | 89.05 |
| Yusmarg | 21 | 33.88 | 74.61 |
| Zanzkar Range | 6, 10 | 33.15 | 76.70 |
| Ziro | 7, 14 | 27.52 | 93.84 |
| Zoji La | 6, 10, 21 | 34.41 | 75.71 |
| Zongla | 179 | 33.69 | 76.95 |
| Zugney | 23 | 27.47 | 90.74 |
| Zuri | 23 | 27.42 | 89.42 |